GUIDELINES FOR DRINKING WATER SAFETY PLANNING FOR WEST BENGAL

DECEMBER 2020

ASIAN DEVELOPMENT BANK

6 ADB Avenue, Mandaluyong City, 1550 Metro Manila, Philippines
Tel +63 2 8632 4444; Fax +63 2 8636 2444
www.adb.org

ISBN 978-92-9262-527-6 (print); 978-92-9262-528-3 (electronic); 978-92-9262-529-0 (ebook)
Publication Stock No. TIM200370-2
DOI: http://dx.doi.org/10.22617/TIM200370-2

Notes:
In this publication, "$" refers to United States dollars.

On the cover: Students of Safique Ahmed Girls High School in Haroa, 24 North Parganas, West Bengal fill their water bottles from the school's taps and hand pump; and Kalpana Manna (inset photo), a resident of Kalicharanpur Village, Purba Medinipur, West Bengal, now has easy access to drinking water (photo by Amit Verma/ADB).

All photos used in this publication are by Amit Verma for ADB and feature project beneficiaries and stakeholders of the ADB-supported West Bengal Drinking Water Sector Improvement Project.

CONTENTS

TABLES AND FIGURE

TABLES

FIGURE

ACKNOWLEDGMENTS

These guidelines for drinking water safety planning were developed for West Bengal by the Public Health Engineering Department (PHED) of the state, with support from the the Asian Development Bank (ADB) and the World Health Organization (WHO), under the West Bengal Drinking Water Sector Improvement Project (WBDWSIP).

Guideline preparation was led and managed by the following individuals and teams:

At the PHED: **Manoj Pant**, principal secretary, PHED; **Debasish Bandyopadhyay**, project director, WBDWSIP; **Animesh Bhattacharya** and **Ajoy Kundu**, chief engineers, PHED; and officials of the WBDWSIP project management and implementation units.

At ADB: **Neeta Pokhrel** and **Luca Di Mario** of the South Asia Urban and Water Division (SAUW), with overall guidance from SAUW Director, **Norio Saito**; inputs from **Saswati G. Beliappa, Saugata Dasgupta**, and **Sourav Majumder**; and assistance from **Elvie Jane Tirano**. **Coral P. Fernandez-Illescas** of the Water Sector Group, Sustainable Development and Climate Change Department (SDCC), served as peer reviewer.

At WHO: **Jennifer De France** and **Kate Medlicott**.

The preparatory team of consultants was led by **Raquel Mendes**, with Prof. **Achinta K. Sengupta**, Prof. **Arunabha Majumder**, and **Hans Enggrob** as members.

ABBREVIATIONS

ADB	Asian Development Bank
BCC	behavior change communication
BIS	Bureau of Indian Standards
COVID-19	coronavirus disease
IWA	International Water Association
MDWS	Ministry of Drinking Water and Sanitation
MLD	million liters per day
NGO	nongovernment organization
PHED	Public Health Engineering Department
SOP	standard operating procedure
SSP	sanitation safety plan
SWM	smart water management
WASH	water, sanitation, and hygiene
WBDWSIP	West Bengal Drinking Water Sector Improvement Project
WHO	World Health Organization
WSP	water safety plan
WSSP	water and sanitation safety plan

The residents of Kalicharanpur are also beneficiaries of the new drinking water project in West Bengal, the West Bengal Drinking Water Sector Improvement Project. ADB and the Government of India signed in November 2018 a $240 million loan and a $5 million grant and technical assistance to provide safe and sustainable drinking water to about 2.6 million people in four districts of the state of West Bengal affected by arsenic, fluoride, and salinity.

1

BACKGROUND

India's population depends largely on groundwater as its major source of drinking water. But in many parts of the country, including West Bengal, groundwater contamination with arsenic, fluoride, and other naturally occurring inorganic contaminants is a major issue. Prolonged exposure to high levels of arsenic in drinking water could lead to serious health conditions such as keratosis, melanosis, and cancer, while high fluoride exposure may result in dental, skeletal, and nonskeletal fluorosis.

At the request of the Government of West Bengal, made through the Government of India, the Asian Development Bank (ADB) is assisting the state government in achieving its Vision 2020 (PHED 2011) of providing safe, reliable, and sustainable drinking water to selected districts in West Bengal. Standards set by the Government of India will be followed.

The West Bengal Drinking Water Sector Improvement Project (WBDWSIP) will benefit about 2.6 million people in selected areas affected by arsenic and fluoride contamination of groundwater and by drinking water salinity in North 24 Parganas, South 24 Parganas, East Medinipur, and Bankura districts in West Bengal.[1]

The project will ensure drinking water security in those areas through 24/7 piped water supply schemes. Surface water uncontaminated with arsenic and fluoride will be the main water source.

The project will construct climate-resilient drinking water infrastructure and strengthen institutions and capacity of stakeholders on drinking water service delivery. Household metered connections will be set up on a district metering area basis through around 4,800 kilometers of water distribution network. Three treatment plants and 110 storage reservoirs will be built—all equipped with smart water management devices and connected to a digitized monitoring system; and a control and command center. The project has also institutionalized an innovative, inclusive, and sustainable service delivery model, wherein a state-level body, the Public Health Engineering Department (PHED), will manage the bulk water

---

[1] Details are available at ADB. 2018. *Report and Recommendation of the President to the Board of Directors: Proposed Loan and Administration of Grant and Technical Assistance Grant to India for the West Bengal Drinking Water Sector Improvement Project*. Manila. https://www.adb.org/sites/default/files/project-documents/49107/49107-006-rrp-en.pdf.

delivery facilities, and the respective *gram panchayats* will manage the distribution network and services within the villages.[2]

For this to happen, major pollution problems for surface water arising from rapid industrialization and consequent urbanization in West Bengal must be dealt with. Pollutants are released from industrial units or urban sewage outlets and from nonpoint sources where contaminants like fertilizers and pesticides mixed with agricultural runoff are discharged. These cause contamination and deterioration of surface water bodies. The government has endeavored to reduce the pollution of rivers and lakes through interception, diversion, and the provision of sewerage and sewage treatment facilities in cities and towns; proper effluent treatment in industries; and efforts to minimize the pollution load of surface runoff. Natural events (like floods, droughts, erosion, and siltation) or climatic changes (like sea-level rise and weather changes) present other challenges for the use of surface water as opposed to groundwater use.

Students of Safique Ahmed Girls High School in Haroa, District 24 North Parganas, West Bengal, filling their water bottles from the school's hand pump.

2 *Gram panchayat* refers to village council—the bottom tier of the three-level *Panchayati Raj* system of the local government in India.

Sujata Maity, a resident of Nandan Nayak Bar Village, East Medinipur, West Bengal, at her house.

ADB is assisting the state government in developing a set of water safety planning and sanitation safety planning guidelines or best practices to enable it to deal capably with such concerns. The PHED of West Bengal should become better able to manage bulk supplies, and the *gram panchayats* to manage household supplies. The ADB assistance is being provided under the WBDWSIP, in close collaboration with the World Health Organization (WHO).

Water safety planning and sanitation safety planning are considered international best practices for assessing and managing public health risks from drinking water supply and sanitation systems. These concerns are most important when dealing with widespread contagion like the coronavirus disease (COVID-19) pandemic. Ensuring water security is the vital and first line of defense.

Besides securing the reliable delivery of safe drinking water in the project areas, the project is expected to broaden the emphasis of water quality management to include the operation and management of water supply and sanitation systems. Institutional structures will be strengthened, and stakeholders at all levels of service delivery will gain increased capacity for sustainable system operation and maintenance and public health improvement. Project standards and guidelines that the state can adopt in its other schemes will also be established.

Bula Ghosh taking water from a hand pump installed at her house in South Adampur, Haroa, District 24 North Parganas, West Bengal.

2

ABOUT THE GUIDELINES

This document offers practical guidance for taking a water safety planning approach to **bulk water supply systems**,[3] such as those operated by the Public Health Engineering Department (PHED) of West Bengal, and to all development and implementation stages of the water delivery service schemes—from design to construction and operation and management.

The PHED is the implementing and executing agency of the West Bengal Drinking Water Sector Improvement Project (WBDWSIP), responsible for project delivery (design and construction). It provides rural water services in the state and bulk water services in its urban areas. The department also operates, maintains, and monitors the bulk water supply up to the boundary of the *gram panchayats*.

The use of these guidelines should assist the PHED in strengthening its current operations and future investments in water safety planning. More specifically, the purpose of this guide is to

- provide practical guidance to the PHED management and staff, as well as other state institutions and stakeholders, in managing health risks in bulk water supply systems;
- sensitize all stakeholders in the state to the importance of water safety planning, and build their capacity to ensure the safety of drinking water by means of this approach;
- enhance WBDWSIP design and sustainability; and
- guide the PHED and other stakeholders in developing other projects, preparing appropriate project designs, maximizing the health benefits of those projects, and ensuring their sustainability.

[3] Bulk water supply systems consist of a set of upstream components that connect the water source to the distribution network (retail system). These systems encompass the abstraction, treatment, elevation, transport, and storage of treated water. Retail systems, with components connected to the end user, can be integrated arrangements using the same system throughout, from supply sourcing to end use.

The present guide also gives some guidance on the importance of human behavior aspects (such as water, sanitation, and hygiene [WASH] practices) and on the need to take cost-effectiveness into account when defining improvement actions.

Water safety planning is the focus of this guide, and providers of bulk water supply services are its main intended users. The links between water safety planning and sanitation safety planning are also briefly explained here. This guide describes the first steps in sanitation safety planning, and indicates whether and how the state government, and the *gram panchayats* in particular, should move to a water and sanitation safety planning approach, since many risks and their management and monitoring within the water supply system will be related to on-site sanitation failures and inadequate drainage facilities in the *gram panchayats*.

Separate water and sanitation safety planning guidelines will be prepared for the *gram panchayats*, with support from the three nongovernment organizations (NGOs) engaged for the project. Those guidelines will have the required details and emphasis for continuous WASH improvement and behavioral change management at the community and *gram panchayat* levels, and will be aimed at institutionalizing practices.

Women's Self-Help Group in South Adampur Village, Block Haroa, Gram Panchayat Haroa, District 24 North Parganas, West Bengal.

The water safety planning guidelines presented here are based on and complemented with the following reference documents:

- WHO guidelines for drinking water quality, fourth edition, incorporating the first addendum (WHO 2017);
- WHO and International Water Association (IWA) safety plan manual for drinking water suppliers (WHO and IWA 2009);
- capacity training program in urban water safety planning, prepared by the South-East Asian Regional Office of WHO (WHO SEARO 2016);
- tool kit for the preparation of a drinking water security plan, developed by the Ministry of Drinking Water and Sanitation (MDWS) of the Government of India, and the Water and Sanitation Program (MDWS and Water and Sanitation Program, 2015);
- WHO sanitation safety planning manual (WHO 2015); and
- water safety plan (WSP) documents of the pilot utilities—the bulk water supply systems in North 24 Parganas and Bankura districts.

It should be pointed out here that while these guidelines are intended primarily for the PHED, they are equally relevant to all other entities with drinking water service delivery and health-related responsibilities in West Bengal, such as the *gram panchayats* and block- and district-level entities in the state.

Moreover, the templates and examples presented here are not meant to be exhaustive or narrowly prescriptive. Rather, these should serve as reference in developing and implementing WSPs tailored to each specific water supply system and to local conditions.

Sulekha Maji, a member of the Self-Help Group in Shreechandrapur, Barsal District Bankura, West Bengal, interacting with ADB project team members.

3

WATER AND SANITATION SAFETY PLANNING

A water safety plan (WSP) is a comprehensive risk assessment and risk management tool that includes all steps in the water supply chain, from catchment to consumer. It addresses the identification of risks that could affect water safety and human health, and helps drinking water suppliers improve and maintain water quality and quantity, and continuity of supply.

By developing a WSP, water suppliers will gain a thorough understanding of their system and be able to take a preventive, rather than reactive, approach to its operation, management, and maintenance. The plan will allow them to identify what could go wrong with the system, how they would know when that happens, what the consequences could be, what is now being done to prevent things from going wrong, and what actions could be taken to prepare for and forestall occurrence.

A WSP will not prevent problems from arising, but it will identify those that can occur and provide an opportunity to deal with the cause before significant problems result. The water supplier must be prepared to face the problems and must deal with them in the face of adversity, with the support of other organizations, when necessary, to maintain water safety.

The use of modern information technology (IT), sensors, and mobile devices provides additional opportunities to monitor, manage, and communicate, e.g., to deal with human behavior aspects through mobile phone interaction. Smart water management (SWM) is, therefore, part of the WBDWSIP.

Like the WSP, the sanitation safety plan (SSP) is a risk-based management tool that includes all steps in the sanitation chain, promoting the safe use and disposal of wastewater and sewage sludge. This plan provides a structure that brings together actors from different sectors to identify health risks from wastewater generation, collection, transport, treatment, reuse, recycling, and disposal, and to agree on needed improvements and regular monitoring.

A significant difference between the water and sanitation safety planning approaches is that while WSPs are confined to the drinking water supply chain, SSPs go beyond the sanitation system and expand to downstream health effects to assure of the safety of sanitation-related products and services. For example, if waste streams are reused or recycled in agriculture to produce a food product, the SSP goes "from toilet to farm to table;" if waste streams are released to the environment, the plan goes "from toilet to environment."

Public health, prevalent WASH and other behavioral practices, safe water supply, and safe sanitation are interrelated. Proper sanitation and wastewater treatment are key challenges in ensuring a healthy environment and safe water.

Many risks within the water supply system will be related to sanitation failures, such as inadequate on-site sanitation, poor drainage and sewage treatment, and system leakage. There are clear benefits to integrating the two approaches into a water and sanitation safety plan (WSSP), as follows:

- assessing and managing risks in a holistic manner to minimize related diseases and safeguard human health,
- increasing the robustness of water and sanitation services by establishing integrated mitigation plans,
- promoting good practices and enhancing environmental protection (management of industrial and agricultural effluents), and

The following web addresses provide useful background reading on water safety plan and sanitation safety plan preparation:

- **Water safety plan resources:** https://www.who.int/water_sanitation_health/water-quality/safety-planning/wsp-publications/en/
- **Sanitation safety plan resources** (manual for safe use and disposal of wastewater, gray water, and excreta): https://www.who.int/water_sanitation_health/publications/ssp-manual/en/

Students of Brij Mohan Tewary Girls High School in Nandigram, Purba Medinipur, West Bengal.

Women's Self-Help Group in Kalicharanpur Village, Purba Medinipur, West Bengal in a meeting with the ADB project team.

- meeting national strategic targets of safe access to water and sanitation for all.

DEVELOPING AND IMPLEMENTING A WATER SAFETY PLAN

A Water Safety Plan (WSP) is a tool that helps water suppliers manage their systems in a systematic and sustainable way.

By following a sequence of activities (Figure), water suppliers can ensure that water

Figure: A Step-by-Step Approach to Water Safety Planning

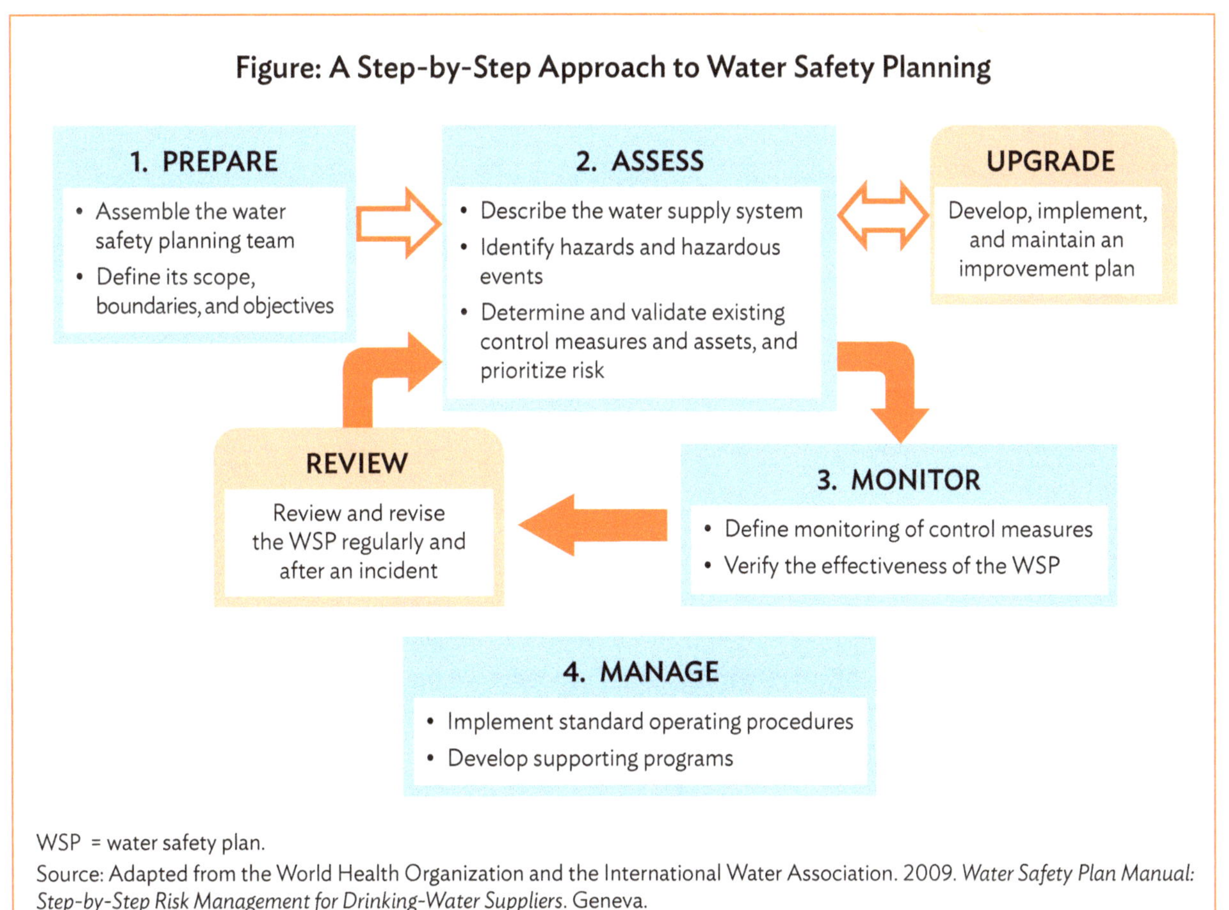

WSP = water safety plan.

Source: Adapted from the World Health Organization and the International Water Association. 2009. *Water Safety Plan Manual: Step-by-Step Risk Management for Drinking-Water Suppliers*. Geneva.

from their systems is safe to drink and does not harm human health.
A WSP covers the four essential water safety planning stages: preparation, assessment, monitoring, and management. To guarantee the success of plan development and implementation, it is important to prepare for water safety planning at the outset and to keep the plan up-to-date by reviewing and upgrading it periodically.

Table 1 presents an example of a broad work plan for the WSP preparation process and the expected output. Each phase is explained in more detail in the following sections.

Table 1: Water Safety Planning—Broad Work Plan and Expected Output

Phase[a]	Description	Output
Phase 1	Preparation	Defined and agreed scope, objectives, and indicators of the plan Trained water safety planning team Engaged stakeholders Water safety planning road map (list of activities, responsibilities, and time frame)
Phase 2	Assessment	System description List of hazards and hazardous events Validated control measures Risk assessment and prioritization Improvement action plan
Phase 3	Monitoring	Operational monitoring program for control measures Verification monitoring/Surveillance program
Phase 4	Management	Designed and implemented SOPs and supporting programs Capacity building/Training program WSP review procedure

WSP = water safety plan, SOP = standard operating procedure.
[a] Each phase is linked to the output of the preceding phase.
Source: Prepared by the consulting team, 2019.

4.1 Phase 1: Preparation

The objective of the preparatory phase is to define the scope and objectives of the WSP, identify the organization that should take the lead in water safety planning, assemble a competent team to drive the process and sustain plan implementation, and gain the support and committed participation of key partner organizations in the process.

At the start of water safety planning, it is also important to prepare a work plan that contemplates not only the development of the plan but also the activities that will ensure its effective implementation. The following questions should be answered in the plan:

- What specific tasks must be done? What outputs should be produced?
- Who will be responsible for these tasks?
- Who will support the performance of the activities, i.e., who else will be involved in this process?
- When will all this be done?

The tasks composing the preparation phase and the output of the phase are presented in Table 2.

4.1.1 Scope and Objectives of the Water Safety Plan

Table 2: Preparation Phase—Key Activities and Expected Outputs

Step No.	Key Activities	Outputs
1.1	Define the scope and objectives of the WSP	Agreed scope and objectives of the WSP Key indicators for evaluating and/or monitoring the benefits and/or impact of the plan
1.2	Assemble a water safety planning team	A multidisciplinary WSP team to develop and implement the plan, with knowledge of the following: - the water supply system and how potential risks to the community water supply, including health, social, environmental, development, and physical planning considerations, can be identified and prioritized; and - smart water management
1.3	Identify relevant stakeholders that must be involved in plan development and implementation	Identified stakeholders and focal points: - with interest in promoting sustained access to safe drinking water, and - able to help mitigate risks

WSP = water safety plan.
Source: Prepared by the consulting team, 2019.

The water safety planning activities start with the identification of the system to be covered in the plan and the **scope** of the plan, i.e., will it apply to the entire water supply system, from water source to point of use, or to only a specific part of the system, such as bulk supply sourcing or distribution to end users?

The overall **objective** and ultimate goal of the WSP should be to safeguard public health. However, implementing such plan can lead to many other positive outcomes, including more available, reliable, and accessible sources of safe drinking water; better water quality; more efficient water supply management and operation; and better communication and collaboration among stakeholders.

For the pilot projects under the West Bengal Drinking Water Sector Improvement Project, it was decided that a water safety plan should be developed for the bulk water supply systems, from the water source up to the boundary of the *gram panchayats* (i.e., up to the ground-level storage reservoir); and for all stages of the development and implementation of the new water delivery schemes, from design and construction to management and operation, including the use of smart water management.

For each defined objective, **indicators** should be established for determining whether and to what extent the objective is being achieved. The indicators should be easily measurable, e.g., through the use of smart water management (SWM) tools, and should allow for evidence of the benefits of implementing the WSP. For each indicator, the water safety planning team must define the goal to be achieved, which can vary over time. Having a way of evaluating the benefits and impact of the plan to confirm its sustainability will enable stakeholders to reflect on the key goals and outcomes of the water safety planning process.

Sample indicators are given in **Appendix 1**.

The objective of developing and implementing a water safety plan for the pilot projects under the West Bengal Drinking Water Sector Improvement Project was to guarantee that proper control measures were taken in all stages of implementation to "ensure that safe water is supplied 24/7, safeguarding public health." This general objective comprises improvements in the following specific areas:

- water quality,
- water accessibility,
- water quantity and continuity,
- capacity building,
- internal and external cooperation,
- security of infrastructure,
- response to emergency situations,
- monitoring and surveillance, and
- sustainability.

Care should be taken not to create new indicators for assessing outcomes but to use monitoring indicators that are already in place or recommended by the World Health Organization for water safety plan impact assessment.

The guidance note on this subject that is being prepared by WHO will be a practical tool for assessing the outcomes and impact of WSP implementation. It will include an indicator framework and data collection forms for field workers.

More information about this tool is provided in Kumpel et al. (2018).

Sandhya Sardar works as cook in the community *anganwadi* (child care and development center) in Baltiya Village, District 24 North Parganas, West Bengal.

4.1.2 The Water Safety Planning Team

The next step in the preparation phase is the formation of a water safety planning team, based on the system to be covered by the WSP and its definition. Each organization in the supply chain (water sourcing, treatment, distribution, and, if possible, end-user premises) and in the various stages of plan implementation (design, construction, and management and operation) should be represented.

The water safety planning team members should have a mix of decision-making capacity, authority, and technical skills. Collectively, they should be able to identify hazards and hazardous events, understand how the risks can be controlled, and bring about the implementation of the necessary improvement actions. The team members will ideally come from diverse backgrounds.

People with one or more of the following characteristics should be considered for membership in the team:

- having knowledge of, and experience in, all aspects of the water supply system, including SWM;
- having the capacity to make spending, staff recruitment, and training decisions, and to decide on changes that should be made in the water supply system to help manage and prevent risks;
- having the ability to take responsibility for design and construction and for the day-to-day management and operation of the water supply system;
- being influential and interested, at both the community level and the administrative level; and
- being familiar with matters related to the links between water and health.

The water safety planning team must appoint a team leader to drive and give focus to the water safety planning effort, and assume responsibility for carrying out the project and ensuring its implementation. This person should have the authority and the organizational and interpersonal skills needed to implement the project.

It is important to divide responsibilities among the team members at the start of the process and to clearly define and record their individual roles.

The table in **Appendix 2** presents the composition, organizational affiliation, and responsibilities of the water safety planning team members for the West Bengal Drinking Water Sector Improvement Project (WBDWSIP) pilot projects.

Under the West Bengal Drinking Water Sector Improvement Project (WBDWSIP):

- The Public Health Engineering Department (PHED) will operate, maintain, and monitor the bulk water supply up to the boundary of the *gram panchayats* (village councils—the lowermost tier of local governance in India), i.e., up to the ground-level service reservoir or the overhead service reservoir. Through the asset management and service delivery framework, the PHED will also regulate the consumer services provided by the *gram panchayats* and assist them with advisory and technical support and training when required.
- The *zilla parsishads* (district councils—the top tier of local governance) and the *panchayat samitis* (block councils—the intermediate level) will provide coordination and technical support, and perform a regulatory and monitoring role at the district and block levels.
- The *gram panchayats* will operate and maintain their respective distribution systems.

In the WBDWSIP pilot projects, the PHED is the leading agency, and the members of the water safety planning teams have been divided into two groups: core and extended members. The core members are responsible for developing, implementing, monitoring, and revising the water safety plan. The extended members are involved only when necessary. They will attend the regular meetings that should take place once the new water delivery systems are operational.

4.1.3 Stakeholders

The process of developing, implementing, and maintaining a WSP, though primarily the responsibility of the lead implementing agency, requires the support and involvement of external **stakeholders**. Different stages of the process will have to involve a different range of stakeholders.

In the case of the WBDWSIP pilot projects, the list of stakeholders is very extensive. Once the key stakeholders were identified, a stakeholder analysis was done to define the required level of interaction.

The water safety planning team must see to it that the planning approach is understood and accepted by everyone associated with water safety within and outside the organizations involved. Relevant capacity-building activities will, therefore, be needed.

External expert(s) could provide the necessary training and technical advice during the project. A good starting point is a project initiation workshop or meeting facilitated by someone with experience in preparing drinking water safety plans. This will give the team members a chance to get to know one another and to understand the various challenges faced by the members. One output of the workshop should be an agreed work plan.

Appendix 3 presents an analysis of stakeholders in the WBDWSIP pilot projects in Bankura and North 24 Parganas districts.

Stakeholders are individuals or groups of people or organizations that are directly or indirectly affected by the water safety plan, may have an interest in the plan or the ability to influence its outcome, either positively or negatively, and could be involved in the implementation of the risk reduction measures.

Over time, the water safety planning team and stakeholders may lose interest in the water safety plan, making it difficult to sustain support for the plan. Officer reassignments and changing roles during implementation can heighten the difficulty.

To help mitigate the risk of waning support for the water safety plan, it is important to

- establish a long-term plan coordination mechanism that will overlap the design-build phase of the project, and system operation and management;
- develop a good implementation plan;
- schedule regular meetings to update and inform everyone concerned about the status of the plan, and to review the plan;
- establish a computerized water safety metric (key performance indicator) through smart water management; and
- achieve good agreement with relevant stakeholders, to gain their commitment to the water safety plan.

Intersector and/or interdepartment coordination at each stage of the project—design, implementation, and operation and maintenance—is needed for a successful water safety plan implementation.

Interdepartment coordination should be done at the state and district levels. District-level coordination committees should meet at least once a month and visit the sites regularly to get a clear picture of the situation. For the pilot projects, an office order was sought from the Government of West Bengal to support the coordination mechanism.

The designated project management unit officer and the water safety planning team should extend their coordination efforts and use the platform of the district development monitoring meetings held monthly at district headquarters. Ongoing development programs are reviewed and discussed at these meetings.

Most of the stakeholders are represented at the district-level steering committee, which also meets monthly. Including water safety planning in this committee's agenda in the near future would allow the matter to be discussed thoroughly at the highest level of the district, and elicit joint action among various stakeholders.

Joint training of officials of the Public Health Engineering Department (PHED), Panchayat & Rural Development, and Health & Family Welfare departments in water supply and sanitation issues, as well as information, education, and communication and awareness-raising activities, must also be provided.

The relevant stakeholders identified in the West Bengal Drinking Water Sector Improvement Project (WBDWSIP) pilot projects are as follows:

- Ministry of Jal Shakti, Department of Drinking Water and Sanitation
- Ministry of Environment, Forest and Climate Change
- Asian Development Bank WBDWSIP project implementation unit and project management unit
- Panchayat & Rural Development Department
- Water and Sanitation Support Organization
- Department of Urban Development and Municipal Affairs
- Department of Agriculture, Cooperation, and Farmers' Welfare
- Department of Environment
- Forest Department
- Department of Health & Family Welfare
- Department of Industry
- Land and Land Reforms Department
- Public Works Department
- Water Resources Investigation & Development Department
- West Bengal Disaster Management Department
- Irrigation and Waterways Department
- Kolkata Port Trust
- West Bengal Housing Infrastructure Development Corporation
- State Environmental Impact Assessment Authority

continued on next page

continued

- State Level Scheme Sanctioning Committee
- State Water and Sanitation Mission
- State Water Investigation Directorate
- District Water and Sanitation Mission
- District Disaster Management Authority/Cell
- West Bengal Pollution Control Board
- West Bengal Police
- West Bengal State Electricity Distribution Company Limited
- East Kolkata Wetlands Management Authority
- *Zilla parishad* (district-level governance)
- *Panchayat samiti* (block-level governance)
- *Gram panchayat* (village-level governance)
- Village water and sanitation committee
- National Environmental Engineering Research Institute (HQ–Nagpur/ Kolkata Divisional Lab)
- National Mission for Clean Ganga
- Fluoride Task Force
- Arsenic Task Force
- Geological Survey of India
- State Laboratory (managed by the PHED)

4.2 Phase 2: Assessment

Whether each of the implementation stages of the water supply system—design, construction, and operation—and the entire water supply chain can deliver water of sufficiently high standards to meet regulatory targets is determined during the risk assessment phase. Included here is the identification of (i) potential hazards, (ii) level of risk, and (iii) appropriate measures for controlling risks.

Identifying and assessing the impact of hazards and hazardous events on the water supply system is a critical component of WSP preparation. The findings will be the basis for developing an improvement plan to ensure that water safety and, hence, public health protection is sustained. A detailed knowledge of the system is required for effective risk assessment. In essence, the risk assessment phase answers the following key questions:

- Which water supply system is used?
- What can go wrong? How? When? Where? Why?
- How serious is the risk of a hazard causing harm?
- How can these hazards be eliminated and their impact minimized?

The tasks and outputs of the risk assessment phase are presented in Table 3.

Hazardous events are events or situations that introduce hazards into, or fail to remove them from, the water supply system.

Hazards are physical, biological, or chemical agents that can cause harm to public health.

Table 3: Risk Assessment Phase—Key Activities and Expected Outputs

Step No.	Key Activities	Outputs
2.1	Describe the water supply system	Validated map and description of the system
2.2	Identify the hazards and hazardous events	Sources of risk to water safety: microbiological, physical, chemical, and water supply disruption (in terms of quantity and continuity of supply) hazards identified along the entire water supply system, from source to tap, for the three implementation phases (design, construction, and operation and maintenance)
2.3	Determine and validate existing control measures; assess and prioritize risk	Identified control measures already existing and their effectiveness Risk matrix for determining the likelihood and severity of risk Risk prioritization
2.4	Develop and implement an improvement plan	Identified additional control measures needed to improve drinking water safety Incremental improvement plan, with prioritized control measures and activities Decisions on where and when each identified improvement will be made, and for whose benefit

Source: Prepared by the consulting team, 2019.

4.2.1 Description of the Water Supply System

The first task of the water safety planning team will be to understand what is already in place. Basically, this involves bringing together available information and knowledge and developing a **detailed, up-to-date description of the water supply system**.

A great deal of information can be recorded and presented in a **map or flow diagram**. Simple pencil-and-paper maps can be prepared. However, the maps should be sufficiently detailed to allow the easy identification of water supply hazards. Flow diagrams created at two levels—system overview (level 1) and water treatment plant details (level 2)—are also helpful. If a geographic information system is available, it should be used in map preparation.

Information gathering, compilation, and validation are not desk activities. They entail infrastructure site visits, as well as discussions with all relevant stakeholders. In many cases, the information is not written down, so knowledge based on experience is valuable and should be taken into account. During data collection, it is important to identify the source and classify the data according to reliability.

Appendix 4 provides guidance on information that must be compiled.

Photographic documentation of the parts of the water supply system during site inspection is necessary to identify and track hazards and hazardous events in the system, and to justify the need to improve or upgrade the facilities.

The water safety planning team should collect sufficient data on relevant quality standards and other legal and regulatory aspects.

Assessing compliance with existing standards and legislation is also part of the water safety planning process. An evaluation of the adequacy of such standards and regulations, e.g., following recent technological developments in the field, should be done as well.

Appendix 5 lists the most relevant legislation, strategy, and guidance documents from the Government of India.

In **Appendix 6**, the bulk supply and distribution systems for the Bankura and North 24 Parganas pilot project areas are described briefly. Existing and new water supply system schematic and/or flow diagrams for the two districts are also presented.

Classroom at Brij Mohan Tewary Girls High School in Nandigram, Purba Medinipur, West Bengal.

4.2.2 Risk Assessment

Risk assessment involves three steps:

- Step 1: Identifying hazards,
- Step 2: Identifying and assessing existing control measures, and
- Step 3: Assessing and prioritizing risks.

Step 1: Identifying Hazards

At each implementation phase and stage of the supply chain, the hazards and hazardous events that may compromise the safety of the water supply system should be identified.

The following questions should be asked: What has gone wrong in the past at this location, what is wrong now, and what could go wrong in the future? How and why might something go wrong? At what times and where? Is anything being done to prevent things from going wrong?

The water safety planning team should consider not only the obvious hazards and hazardous events associated with the water supply, but also the potential occurrence of hazards and hazardous events because of the following:

- technical defects associated with the design and construction phase (e.g., improper choice of water treatment due to inadequate water quality data, defective infrastructure due to improper testing during construction);
- lack of understanding of the water supply system and how it operates;
- improper procedures or malpractices in system management and operation (e.g., irregular maintenance, operator error, inadequate filter washing);
- operational failures (e.g., power shutdown, treatment failure including those due to equipment breakdown or operator error);
- accidental contamination (e.g., chemical spill near water source);
- natural hazard events (e.g., heavy rainfall, landslide, flood, drought);
- changes in or around the system (e.g., salinity intrusion, erosion or siltation around the intake, land use change, construction, new industry);
- climate change (e.g., frequent flooding, drought, storm surge, sea-level rise, causing changes in raw water availability); or
- human-made disasters resulting from neglect or sabotage.

To identify hazardous events, field visits and site inspections and a desktop review of system diagrams should be done, the operational team and members of the various organizational units of the entity should be interviewed, and meetings should be held with stakeholders. The hazard identification should also be based on an analysis and evaluation of past events and information, as well as on studies (e.g., hydrogeological studies, hydraulic modeling, chlorine decay studies).

Hazardous event identification at the pilot project sites for the West Bengal Drinking Water Sector Improvement Project (WBDWSIP) (Appendix 7) brought out the following points:

- Inequitable distribution of water across villages is a major issue, since water distribution across villages is not normally regulated. Water flow at the tail end of the village, and not just closer to the pumping station, water source, or overhead reservoir, must be ensured. This matter needs to be taken care of at the design stage and during water tariff setting to reduce misuse and maintain equitable distribution. A grievance cell, along with community consultation mechanisms, must also be introduced to improve vigilance and maintenance of water distribution, and thus prevent water misuse.
- As water availability is generally low at present, the villagers are not used to high consumption. But even with household water consumption at 70 liters per capita per day, there is every possibility that drainage problems will occur on a large scale in the villages unless villagers start kitchen gardens and use soak pits to hold gray water. If this is not done, vector-borne diseases will increase, cesspools will be created, and roads within the village will be damaged.
- Observing and understanding human behavior patterns is important in managing water supply systems. Wide use of apps on mobile phones is a feature of smart water management that allows more direct and closer interaction with end-consumers to improve water services. Integrated mitigation planning and more direct interaction with end-consumers through mobile devices will make the water (and sanitation) systems more robust.
- The WBDWSIP proposal suggests building a long pipeline network through remote areas and, in some cases, even through hilly tracks. Exposed pipelines could get damaged, intentionally or otherwise. Some vigilance mechanism must be established to keep the network safe and prevent water theft. Introducing a smart water management system will help eliminate illegal water connections.

For the WBDWSIP pilot projects, hazards were categorized into four groups, namely:

- **Microbiological (M) contamination**, referring to microorganisms with the potential for acute impact on public health, such as bacteria, viruses, and pathogenic protozoa.
- **Chemical (C) contamination**, referring to chemicals that could be present at levels resulting in acute or chronic health effects, including chemicals with accumulative properties such as metals and organic substances, and pharmaceuticals.
- **Physical (P) impact**, referring to constituents with aesthetic impact, such as those affecting the clarity, color, taste, and odor of water. Aesthetic concerns, while not directly health-related, can have an important impact on overall water safety in a community. For example, water that is safe but has a bad appearance, taste, or odor may not be accepted by consumers, who may choose alternatives that are aesthetically acceptable but also less safe. Water that tastes good, on the other hand, has a positive impact on

continued on next page

continued

people's general feeling of well-being and potentially on the overall vitality and sustainability of the community.

- **Water shortage or water supply interruption** due to operational failure, human-made disasters, natural hazardous events, or climatic changes (e.g., sea-level rise, causing elevated salinity upstream).

When describing hazardous events, be specific. Clearly indicate what hazard could be introduced, and how.

A good hazardous event statement reads like this: X happens (to the water supply) because of Y.

X = what can happen to the water supply; Y = how it can happen (i.e., cause)

Table 4 lists some hazardous events identified at the pilot project locations in Bankura and North 24 Parganas districts.

Rinku Somanto runs a tailoring shop in South Adampur Village, Haroa, District 24 North Parganas, West Bengal.

Table 4: Some Hazardous Events Identified in the Pilot Projects

Implementation Phase	Process Step	Component	Hazardous Event	
			X = What Can Happen to the Water Supply	Y = How It Can Happen (i.e., cause)
Design	Water source	Intake	Nonavailability or shortage of raw water	Improper choice of water source due to flaws in initial bathymetric survey or unattended long-term morphological changes (e.g., ongoing bank erosion, bank accretion, high sediment concentration), followed by incorrect placement of intake point
Design	Water source	Intake	Nonavailability of raw water	Long-term changes in water level statistics at the river intake causing flooding, drought, or elevated salinity, due to inadequate hydrological or coastal studies or failure to factor in climate change
Design	Treatment	Water treatment plant	Inadequate water treatment	Improper design of water treatment process due to - lack of detailed assessment of raw water quality, or - failure to consider seasonal variations in raw water quality
Design	Transport	Pumping station	Water supply interruption	Pumping station malfunction due to wrong choice of pump capacity and insufficient number of pumps
Design	Transport	Distribution network	Water supply interruption	Leaks or bursts in transmission main or pipeline due to insufficient number of valves and thrust blocks Wrong choice of pipe material
Design	Transport	Transmission main, distribution network	Water contamination	Wrong choice of pipe material
Design	Transport	Transmission main, distribution network	High energy consumption	Poor hydraulic design
Design	Storage	Service reservoirs	Water contamination	Failure to consider proper safety measures to prevent animal intrusion
Design	General	General	Water supply interruption	Improper design due to lack of reliable primary and secondary data in survey report
Construction	Water source	Monitoring and surveillance	Late completion of targeted project	Lack of monitoring and surveillance during construction

continued on next page

Table 4 *continued*

Implementation Phase	Process Step	Component	Hazardous Event	
			X = What Can Happen to the Water Supply	Y = How It Can Happen (i.e., cause)
Construction	Treatment	Water treatment plant	Supply of untreated or contaminated water	Faulty construction of components of the proposed treatment system
Construction	Transport	Distribution network	Pipeline leaks and/or bursts	Technical fault due to contractor's failure to meet the technical specifications for the pipelines (poor joints, inadequate depth)
Construction	Storage	Service reservoirs	Structural damage	Due to - poor-quality construction materials - improper soil testing leading to cracks in structure
Construction	General	General	Delay in construction	Due to - lack of monitoring and surveillance during construction - poor workmanship, improper work methods, technical disputes - improper choice of agency and inadequate control over subcontracting - nonavailability of materials at construction site - local labor unrest
Construction	General	General	Water contamination	Improper cleaning of reservoirs and transmission mains before commissioning
Operation	Water source	Intake	Water contamination	Further drainage of water treatment sludge to the intake point
Operation	Water source	Intake	Water contamination	Runoff from farm fields
Operation	Water source	Intake	Water supply interruption	Improper maintenance of pumping station
Operation	Water source	Intake	Water supply interruption	Improper functioning of electromechanical components due to nonavailability of trained and dedicated technical support
Operation	Water source	Intake	Water supply interruption	Sudden power outage
Operation	Water source	Intake	Water supply interruption	Irregular pumping and/or supply of raw water due to physical damage in intake well
Operation	Water source	Intake	Water supply interruption	Nonavailability of raw water due to silting (sedimentation) in dam

continued on next page

Table 4 *continued*

Implementation Phase	Process Step	Component	Hazardous Event	
			X = What Can Happen to the Water Supply	Y = How It Can Happen (i.e., cause)
Operation	Water source	Intake	Water supply interruption at certain times	Elevated salinity levels, which depend on tide and river flow (pumping can be done only during low tide and neap tide, and outside the dry season)
Operation	Water source	Intake	Water shortage	Due to lack of rainfall in a particular year
Operation	Treatment	Disinfection	Supply of untreated or contaminated water	Inadequate treatment due to the presence of unmonitored residual chlorine in the water supply
Operation	Treatment	General	Supply of untreated or contaminated water	Inadequate water treatment due to operators' lack of proper technical knowledge
Operation	Treatment	General	Supply of untreated or contaminated water	Inadequate water treatment due to malfunctioning of treatment processes (oxidation, coagulation and flocculation, filtration, sedimentation, disinfection) caused by inadequate O&M
Operation	Treatment	General	Water supply interruption	Power failure
Operation	Treatment	General	Supply of untreated water	Inadequate treatment due to contractor's failure to provide qualified O&M personnel
Operation	Treatment	General	Supply of untreated water	Inadequate treatment due to nonavailability, or low quality and irregular supply, of chemicals
Operation	Treatment	General	Supply of untreated water	Inadequate treatment due to lack of reliable water quality data
Operation	Transport	Distribution network	Water supply interruption	Unskilled manpower, lack of knowledge needed to operate valves
Operation	Transport	Transmission main, distribution network	Water supply interruption	Lack of coordination among contractors
Operation	Transport	Distribution network	Water contamination	Ingress of foreign matter following pipe bursts or accidental breakage, and as a result of leakage
Operation	Transport	Transmission main, distribution network	Water supply interruption	Leakage or pipe bursts
Operation	Transport	Transmission main, distribution network	Water contamination	Unauthorized connections and potential backflow from illegal tapping

continued on next page

Table 4 *continued*

Implementation Phase	Process Step	Component	Hazardous Event	
			X = What Can Happen to the Water Supply	Y = How It Can Happen (i.e., cause)
Operation	Transport	Transmission main, distribution network	Water contamination	Biofilm formation due to lack of chlorine
Operation	Storage	Service reservoirs	Water contamination	Accumulation of solid particles on the reservoir floor and creation of breeding ground of algae, etc., due to lack of, or improper and sporadic cleaning
Operation	Storage	Service reservoirs	Water supply interruption	Over withdrawal of water by consumers nearer to the source
Operation	Storage	Capacity of buffer reservoirs	Interruption of water supply	Water intake pumping at certain intervals during the tide (daily), during the neap–spring cycle (biweekly), and during the season, due to elevated salinity or low freshwater availability, requiring buffer reservoir storage capacity
Operation	Storage	Service reservoirs	Water contamination	Intrusion and ingress of vermin, reptiles, and birds through improperly covered inspection openings or through openings in the roof and/or air vents
Operation	Storage	Service reservoirs	Water contamination	Scaling and formation of biofilm on the walls
Operation	Storage	Service reservoirs	Water contamination	Microbial growth due to high detention periods
Operation	General	General	Supply of untreated or contaminated water at any place along the water supply chain from source to tap	Lack or inadequacy of water quality testing
Operation	General	General	Water contamination	Security breaches
Operation	General	General	Water supply interruption at any place along the water supply chain from source to tap	Nonavailability of spare parts in times of crisis or other unpredictable events
Operation	General	General	Water supply interruption at any place along the water supply chain from source to tap	SCADA failure due to disruption in online communication or lack of robust SCADA design (no automatic fallback when manual controls fail)
Operation	General	General	Water contamination and water supply interruption at any place along the water supply chain from source to tap	Occurrence of natural disasters (floods, earthquakes, droughts, etc.)

continued on next page

Table 4 *continued*

Implementation Phase	Process Step	Component	Hazardous Event	
			X = What Can Happen to the Water Supply	Y = How It Can Happen (i.e., cause)
Operation	General	General	Water contamination and water supply interruption at any place along the water supply chain from source to tap	Improper management of available resources Transfer of trained personnel Transfer of management personnel
Operation	General	General	Water contamination and water supply interruption at any place along the water supply chain from source to tap	Security breaches

O&M = operation and maintenance, SCADA = supervisory control and data acquisition.
Source: Prepared by the consulting team, 2019.

Step 2: Identifying and Assessing Existing Control Measures

For each hazardous event identified, the existence of control measures, and what those are, must be determined. It is very important to verify whether such restrictions are effective in eliminating or reducing the identified risks; their proper operation should not be taken for granted.

If the control measures are ineffective, or if none are currently in place for an identified significant risk, this matter should be noted, together with suggestions for improvement. Table 5 presents some control measures used in the WBDWSIP pilot projects.

Control measures (also referred to as barriers or mitigation measures) are any activities and processes that can be used to prevent, eliminate, or significantly reduce the occurrence of a water safety hazard.

Students of Brij Mohan Tewary Girls High School in Nandigram, Purba Medinipur, West Bengal.

Table 5: Some Existing Control Measures in the Pilot Projects

Implementation Phase	Component	Hazardous Event	Control Measures
Design	Intake	Nonavailability of raw water due to flooding and subsequent malfunctioning of intake facilities	Use of flood forecasting and early warning system, and provision for the use of alternative water source in case of supply disruption
Design	Intake	Nonavailability of raw water due to elevated salinity	Use of salinity forecasting for SWM of pumping, and use of buffer reservoir storage capacity or other supplementary water sources
Design	Water treatment plant	Inadequate water treatment due to improper design of treatment process, which - lacks a detailed assessment of raw water quality, and - does not consider seasonal variations in raw water quality	Treatment plant design based on available raw water quality data and survey data to effectively remove all contaminants Conventional type of water treatment plant has been proposed
Design	Pumping station	Water supply interruption caused by malfunctioning of pumping station due to wrong choice of pump capacity and insufficient number of pumps	Proof checking to be vetted by third party, such as the design and supervision consultants, and approved by the Public Health Engineering Department
Design	Distribution network	Water supply interruption following a transmission main or pipeline leakage and/or burst due to insufficient number of valves and thrust blocks Wrong choice of pipe material	Pipeline design based on detailed route survey; provision of required number of valves Use of proper laying and jointing procedures as described in bid documents
Design	Transmission main, distribution network	Water contamination due to wrong choice of pipe material	Conformity of pipe material to prevailing BIS and Central Public Health and Environmental Engineering Organization specifications to be thoroughly checked during procurement
Design	Transmission main, distribution network	High energy consumption due to poor hydraulic design	Adequate design and preliminary energy audit during the design phase (based on equipment specifications)
Design	Service reservoirs	Water contamination caused by failure to consider proper safety measures against animal intrusion	Property boundary wall along with mild steel or similar gate has been considered to prevent animal intrusion
Design	General	Water supply interruption caused by improper design due to lack and unreliability of primary and secondary data in survey report	Third-party checking of survey report
Design	General	Water supply interruption caused by elevated salinity levels at certain times of the day/month/year	Establishment of buffer reservoir storage, where capacity matches the time period during which pumping is disrupted Establishment of other supplementary water sources

continued on next page

Table 5 *continued*

Implementation Phase	Component	Hazardous Event	Control Measures
Construction	Water treatment plant	Supply of untreated or contaminated water due to faulty construction of components of the proposed treatment system	Standard project requirements specified in the bid document Skilled personnel to be involved in implementation Strict monitoring to be done during implementation
Construction	Distribution network	Pipeline leakage and/or burst caused by technical fault due to contractor failure to meet the technical specifications for the pipelines (substandard joints, inadequate depth)	Pipe laying and jointing standards addressed in the tender document Quality assurance and control action plans have been considered
Construction	Service reservoirs	Structural damage due to - poor-quality construction materials - improper soil testing leading to cracks in structure	Third-party checking of soil testing report Proper sourcing of materials under IS 383 (coarse and fine aggregates for concrete) and other codel provisions, and high-quality monitoring of various tests at site level Leakage testing to be done in the field after completion of work under IS 3370-4 (concrete structures for storage of liquids) and other relevant standards
Construction	General	Delay in construction due to - lack of monitoring and surveillance during construction - poor workmanship, improper work methods, technical disputes - improper choice of agency and inadequate control over subcontracting - nonavailability of materials at construction site - local labor unrest	Periodic monitoring to be conducted to ensure that construction proceeds as scheduled Inclusion of clause for delay damage in the tender document has been considered
Construction	General	Water contamination caused by improper cleaning of reservoirs and transmission mains before commissioning	Provision in the tender document and proper supervision of works will be carried out
Operation	Intake	Water supply interruption due to improper maintenance of pumping station	Maintenance plan to be carried out
Operation	Intake	Water supply interruption caused by improper functioning of electromechanical components due to nonavailability of trained and dedicated personnel	Specific training program for electromechanical staff

continued on next page

Table 5 *continued*

Implementation Phase	Component	Hazardous Event	Control Measures
Operation	Intake	Water supply interruption caused by irregular pumping and/or supply of raw water due to physical damage in intake well	Regular cleaning of suction pipe screen at intake and liaison with the authority concerned Ganga Action Plan, Namami Ganga Program
Operation	Intake	Water supply interruption caused by elevated salinity levels at certain times of the day/month/year	Intelligent pumping at the intake during low/no-salinity periods using SWM (sensors, real-time monitoring), or shift to other supplementary water sources
Operation	General	Supply of untreated or contaminated water caused by inadequate water treatment due to lack of proper technical knowledge of operators	Periodic refresher training is provided
Operation	General	Water supply interruption due to power failure	Backup power facility is available
Operation	General	Supply of untreated water because of inadequate treatment following the nonavailability or low quality and irregular supply of chemicals	Chemicals procured have testing certificates conforming to BIS requirements Stock management in place
Operation	Distribution network	Water supply interruption due to - unskilled manpower - lack of proper knowledge of valve operation	Capacity-building plan to be developed and implemented to enhance operator capacity
Operation	Transmission main, distribution network	Water contamination caused by unauthorized connections and potential backflow from illegal tapping of water	Involvement of *gram panchayat* (village council) and village water and sanitation committee in controlling unauthorized connections
Operation	Service reservoirs	Water contamination caused by the accumulation of solid particles on the reservoir floor and creation of breeding ground of algae, etc., due to lack of, or improper and sporadic cleaning	Cleaning and disinfection and maintenance plan to be carried out
Operation	General	Water contamination and water supply interruption due to natural disasters (flood, earthquake, drought, etc.)	Rules and regulations for disaster management exist Forecasting and early warning systems to be established and emergency plan to be developed
Operation	General	Water contamination and water supply interruption due to - improper resource management - transfer of trained personnel - transfer of management personnel	SOPs and other guidelines have been considered

BIS = Bureau of Indian Standards, SOP = standard operating procedure, SWM = smart water management.
Source: Prepared by the consulting team, 2019.

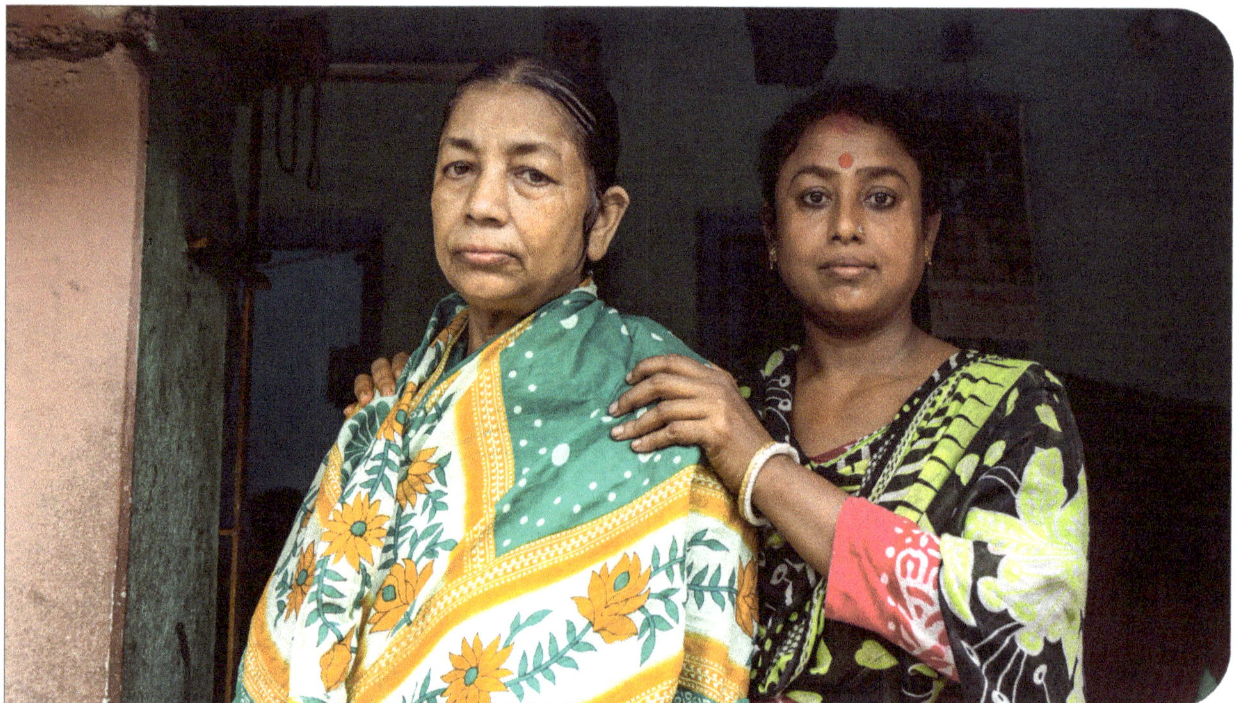

Bula Ghosh with her mother-in-law, Madhurani Ghosh, at her home in South Adampur Village, Haroa, District 24 North Parganas, West Bengal.

Step 3 : Assessing and Prioritizing Risks

Risk assessments can be done in several ways. Depending on the scale and complexity of the water supply system, the water safety planning team may decide to carry out a descriptive assessment of the risks, or a more rigorous semiquantitative risk assessment, where the team assesses the likelihood of the hazardous event and the consequence or severity of its impact, using a probability and severity matrix to determine the level of risk. Computerizing such risk matrices and linking these to a metric as part of smart water management (SWM) could embed the risk assessment in daily overall monitoring.

Generally, it is better for the water safety planning team to start with less complicated risk assessments and progress to more precise approaches as information, skills, and resources become increasingly available.

Examples of risk matrices can be found in the guidelines for drinking water quality (WHO, 2017).

To allow for full ownership of the risk assessment, the process should be validated by all relevant stakeholders.

Appendix 7 summarizes the risk assessment done for the bulk water supply system pilot projects in Bankura and North 24 Parganas districts.

Risk assessment involves identifying the threats to human health posed by each hazard, in relation to both the likelihood and the severity of occurrence.

Risk refers to the likelihood of a hazard causing harm to exposed populations within a specific time frame, and the magnitude or consequences of that harm. The following formula is used to calculate risk:

Risk = Likelihood x Consequence (Severity)

In water safety planning for the pilot projects under the West Bengal Drinking Water Sector Improvement Project, risk assessment involved two tasks: (i) developing a risk matrix and defining the likelihood and consequence of risk, and (ii) assigning likelihood and severity to each identified risk and mapping the results in the matrix to obtain a risk ranking.

For the first task, the water safety planning team drew up the risk matrix and definitions for the likelihood (unlikely, possible, very likely, or almost certain) and consequence (minor, moderate, major, or catastrophic impact) categories. Semiquantitative risk assessment was used, and two risk matrices—a 3x3 matrix for the design and construction phases and a 5x5 matrix for the operation phase—were defined.

The aim was to distinguish significant risks from less significant risks. Risks were categorized as follows:

- low (clearly not a priority, since control measures seem to be adequate);
- medium (medium- or long-term priority, and requiring some improvement in control measures in the medium or long term); and
- high (clearly a priority, and requiring urgent attention and urgent improvement in control measures).

The risk matrices below show how the risk ratings were used to evaluate quantitatively the hazards in the system. To give immediate visibility to the level of risk, the scores are displayed on a red/yellow/green scale in the tables. The two other tables summarize the definitions of the likelihood and severity for the design and construction phase and the operation phases, respectively.

Risk Matrix for the Ranking of Risks in the Design and Construction Phase (3x3 matrix)

			CONSEQUENCE		
			No/Minor Impact	Moderate Impact	Major Impact
			1	2	3
LIKELIHOOD	Unlikely	1	1	2	3
	Possible	2	2	4	6
	Most likely	3	3	6	9

Risk score	≤2	3–5	≥6
Risk level	Low	Medium	High

continued on next page

continued

Suggested Descriptions for the Likelihood and Consequence (Severity) of Risk in the Design and Construction Phase (3x3 matrix)

Likelihood		Definition (Design and Construction Phase)
1	Unlikely	Highly improbable that it will happen
2	Possible	This is possible and could happen under certain circumstances
3	Most likely	Has the potential to happen

Consequence		Definition (Design and Construction Phase)
1	No/Minor impact	Wholesome water (quality) Design quantity criteria is met (70 lpcd) (quantity) Design coverage criteria of project area is met (100%)
2	Moderate	Aesthetic or noncompliance issue, not health related (quality) Design quantity criteria has less than 15% error (at least 60 lpcd will be met [quantity] or at least 85% coverage area is met)
3	Major	Potential immediate health effect (quality) Design quantity criteria has 15%–25% error (at least 45 lpcd will be met [quantity] or at least 75% coverage area is met)

lpcd = liter per capita per day.
Source: Prepared by the consulting team, 2019.

Risk Matrix for the Ranking of Risks in the Operation Phase (5x5 matrix)

			CONSEQUENCE				
			Insignificant	Minor	Moderate	Major	Catastrophic
			1	2	3	4	5
LIKELIHOOD	Most unlikely	1	1	2	3	4	5
	Unlikely	2	2	4	6	8	10
	Possible	3	3	6	9	12	15
	Very likely	4	4	8	12	16	20
	Almost certain	5	5	10	15	20	25

Risk score	≤5	6–14	≥15
Risk level	Low	Medium	High

Source: Prepared by the consulting team, 2019.

continued on next page

continued

Suggested Descriptions for the Likelihood and Consequence (Severity) of Risk in the Operation Phase (5x5 matrix)

Likelihood		Definition (Operation Phase)
1	Most unlikely	Highly improbable that it will happen
2	Unlikely	This is possible and cannot be ruled out completely
3	Possible	This is possible and could happen under certain circumstances
4	Very likely	This has the potential to happen
5	Almost certain	This is expected to happen

Consequence		Definition (Operation Phase)
1	Insignificant	Wholesome water (quality) Minimum of 50 lpcd (quantity) or 2-hour disruption in water supply (continuity)
2	Minor	Short-term or localized noncompliance or aesthetic issue, not health related (quality) At least 45 lpcd (quantity) or 6-hour disruption in water supply (continuity)
3	Moderate	Widespread aesthetic or long-term noncompliance issue, not health related (quality) At least 30 lpcd (quantity) or 12-hour disruption in water supply (continuity)
4	Major	Potential immediate health effect (quality) At least 10 lpcd (quantity) or 24-hour disruption in water supply (continuity)
5	Catastrophic	Potential illness and long-term health effects (quality) Less than 10 lpcd (quantity) or more than 1 day sudden (not planned) disruption in water supply (continuity)

lpcd = liter per capita per day.
Source: Prepared by the consulting team, 2019.

For the second task, the water safety planning team carried out risk scoring following site visits, examination of data, and training workshops with local staff and relevant stakeholders. When the effectiveness of the control measure was not known, the risk was calculated as though there were no control measures in place.

The assignment of likelihood was essentially based on staff experience. The questions considered in assessing likelihood were: Has this happened before in the system? If so, how often does this event occur? Is it likely to happen again in the foreseeable future? Are there weaknesses in the system that will allow the failure to occur more often in the future than in the past? The likelihood of occurrence was assessed on the basis of the answers to these questions.

The degree of severity of the hazards was based on information, as agreed on at the start of the process, using World Health Organization guidelines.

Lack of information and the number of stakeholders involved made it a challenge to quantify the likelihood and consequences of several potential risks.

4.2.3 Improvement Plan

A key output of water safety planning is the improvement action plan addressing key risks and other concerns identified in the risk assessment.

In the previous task, the water safety planning team determined whether new or improved control measures must be developed to improve risk management. Those measures, together with the person or agency responsible for each one, must be identified in the improvement plan so they can be implemented and managed.

Improvement actions can be capital works, operational measures (including external service acquisition), behavioral measures, or a combination of the three. The water safety planning team should take the following factors into account when evaluating different control measures:

- potential for improving existing controls;
- technical effectiveness;
- acceptability and reliability in relation to local cultural and behavioral practices;
- implementation, management, and monitoring responsibilities; and
- training, communication, consultation, and reporting requirements.

The cost of the proposed control measures and the timing of their implementation must also be stated in the plan. Some improvements or control measures can be implemented immediately, at little or no cost. Others, such as more use of SWM, will have to be implemented in stages and may require a substantial budget and additional external resources. Some risks that may be too costly to be fully dealt with during the plan period may have to remain

Sutapa Midya, principal at Brij Mohan Tewary Girls High School in Nandigram, Purba Medinipur, West Bengal.

substantially unaddressed for the time being, although their occurrence is an acknowledged possibility.

A computerized key performance indicator metric for the effectiveness of operation and maintenance, developed over time, will help make action plans more cost-effective.

It is best not to try to do everything at once but to work toward water quality targets or objectives through incremental improvements. These can be planned within feasible and realistic time frames (e.g., 3–5 years) using the water safety planning approach, and prioritized on the basis of the risk category of the hazard each control measure is designed to prevent or minimize.

The improvement plan should be documented and shared with all those responsible for the improvement measures. Its implementation status should be monitored and recorded by the water safety planning team to ensure that every planned action is taken.

Various aspects of the improvement plans for the West Bengal Drinking Water Sector Improvement Project (WBDWSIP) pilot projects are summarized in **Appendix 8**. Both plans are based on risk ratings, and assign priority to hazards with high ratings.

New control measures may sometimes introduce new risks that must be addressed by the water safety planning team.

Chlorinating water, for example, may induce some consumers, unhappy with the new taste, to start taking water from alternative, potentially unsafe, sources. Community education in taste perception, provided together with the launch of the new measure, will be vital in securing a more positive response from consumers.

4.3 Phase 3: Monitoring

Monitoring (Table 6) is a key requirement of the water safety planning process. It reveals the extent to which the control measures are being implemented and are effective, and shows how well the water safety planning process as a whole is able to ensure the availability of safe drinking water (performance verification). The monitoring results can serve as evidence to support the rightness of the new measures. On the other hand, they can also be employed to justify doing things the old way.

Extensive use of new sensing and monitoring technologies emerging from the Internet-of-Things era provides excellent opportunities for enhanced control and monitoring of critical parameters. A computerized water safety metric for SWM can be gradually built up in the organization to improve water safety. It can make water resilience an integral part of daily overall monitoring, with results accessible throughout the system and possibly capable of triggering automated actions.

Table 6: Monitoring Phase—Key Activities and Expected Outputs

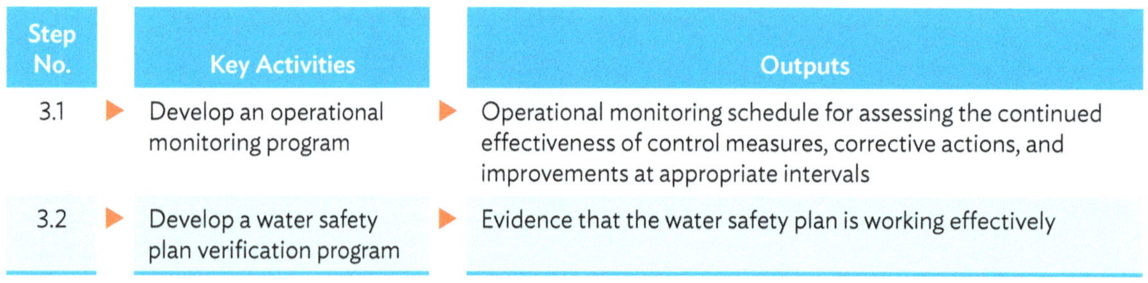

Step No.	Key Activities	Outputs
3.1	Develop an operational monitoring program	Operational monitoring schedule for assessing the continued effectiveness of control measures, corrective actions, and improvements at appropriate intervals
3.2	Develop a water safety plan verification program	Evidence that the water safety plan is working effectively

Source: Prepared by the consulting team, 2019.

4.3.1 Operational Monitoring

Operational monitoring involves checking control measures to provide real-time feedback on whether they are working properly. It involves quick measures: setting target levels and action limits for each of the controls, and identifying the necessary corrections that must be made as quickly as possible if an action limit is breached.

Appendix 9 presents features of the operational monitoring plan prepared by the water safety planning teams for the Bankura and North 24 Parganas pilot projects.

Answering the following questions will help the water safety planning team develop operational monitoring programs for control measures and associated schedules:

- Why monitor and/or inspect X?
- What does monitoring and/or inspection of X require?
- How will X be monitored and/or inspected?
- When and where will X be monitored and/or inspected?
- Who will monitor and/or inspect X?
- What is the acceptable range of values for X? (These can be quantitative or qualitative, i.e., yes/partially/no.)
- What corrective actions will be taken when X is outside the target range?
- Who will carry out the corrective actions?
- What records and reporting are expected from X?
- What training is needed to implement the operational monitoring program for X (training for individuals responsible for sampling, testing, and analysis)?

Source: World Health Organization. 2012. *Water Safety Planning for Small Community Water Supplies: Step-by-Step Risk Management Guidance for Drinking-Water Supplies in Small Communities*. Geneva.

Members of the Women's Self-Help Group in Baltiya Village, District 24 North Parganas, West Bengal.

4.3.2 Verification Monitoring

Verification monitoring involves three activities done together to provide evidence that the water safety plan (WSP) process is working effectively. They are as follows:

- compliance monitoring,
- internal and external auditing, and
- consumer satisfaction survey.

Compliance monitoring is typically based on water quality testing, and the results are checked against established national water quality standards. Monitoring is usually done by another organization not involved in the day-to-day operation of the water supply system, such as a public health officer or inspector, and is focused on the endpoints of the system.

Internal and external auditing helps maintain the quality of WSP implementation. Auditing ensures that water safety planning is still having a positive outcome by checking the quality and effectiveness of the process. It examines the activities enumerated during the planning and evaluates whether they are being carried out in practice and records are kept where required.

The external audit team may include government officials or the regulatory authority, or water quality experts from neighboring larger utilities.

The following questions should be considered during the audit:

- Have all significant hazards and hazardous events been identified?
- Have appropriate control measures been included?
- Have appropriate operational monitoring procedures been established?
- Have appropriate operational or critical limits been defined?
- Have corrective actions been identified?
- Have appropriate verification monitoring procedures been established?
- Have hazardous events with the most potential to cause problems for human health been identified, and has appropriate action been taken?

Consumer use of, and satisfaction with, the water supply system is an important indicator of whether the system is operating effectively. In the case of bulk supply systems, the consumer is the *gram panchayat*.

More information about the development and implementation of water safety plan auditing schemes is provided in WHO and IWA (2015).

For the pilot projects under the West Bengal Drinking Water Sector Improvement Project, a surveillance program needs to be developed for the success of water safety plan implementation, and direct links need to be established between the Public Health Engineering Department and the consumer, such as a "consumer care center" to receive consumer complaints and/or suggestions related to public water supply.

For proper surveillance, the water safety planning team suggested the inclusion of the following agencies: village water and sanitation committees, *gram panchayats* (village councils), the block health department, the Water and Sanitation Support Organization, and the Public Health Engineering Department.

A sanitary survey of drinking water sources and the use of field kits should be encouraged to support water quality monitoring. Crucial to the successful implementation of a surveillance mechanism is the strengthening of VWSCs, Integrated Child Development Services, health workers under the Accredited Social Health Activist Program, and nongovernment organizations through training and the assignment of greater responsibilities, and the involvement of women members.

All operational monitoring and verification data should be documented, filed, and shared with relevant stakeholders. There may be legal or other requirements involving the submission of reports to public health or regulatory officials. Over time, this documentation will be helpful as results are analyzed to explain historical performance and occurrences and to show what risks occur and with what frequency. This information will help in improving the continued implementation of the water safety plan, especially in justifying investments.

4.4 Phase 4: Management

The preparation and updating of the WSP should be viewed not as a one-time undertaking but as an integral part of the ongoing, day-to-day operation, maintenance, and management of the water supply system, with a view to ensuring its sustainability into the future in terms of financial resources and the natural resource base. Management procedures that describe actions taken during normal and incident operating conditions and a scheduled review of the water safety planning tasks and output of the management phase are discussed below (Tables 7 and 8).

Table 7: Management Phase—Key Activities and Expected Outputs

Step No.	Key Activities	Outputs
4.1	Develop and implement SOPs and supporting programs	SOPs for standard (normal) and emergency situations, shared with all members of the water safety planning team and operators responsible for managing the water supply List of supporting activities needed and available well-established record-keeping and documentation system, with transparent communication procedures
4.2	Develop procedure for WSP review and updating	Method of reviewing the WSP periodically

SOP = standard operating procedure, WSP = water safety plan.
Source: Prepared by the consulting team, 2019.

4.4.1 Management Procedures

Standard Operating Procedures

All systems require instructions for their operation and management. **Standard operating procedures (SOPs)** are written instructions that describe the steps or actions to be taken during normal operating and maintenance conditions, operational monitoring actions, and corrective actions that must be resorted to when operational monitoring parameters exceed operating limits. SOPs also describe safety issues, frequency of task execution, and recording.

Staff must know about these procedures and how they can be accessed when necessary. Training in the new or revised procedures should be provided. The management unit responsible should also establish a means of checking to make sure that the procedures and training indeed result in an effective system for the supply of safe drinking water.

Table 8 presents a list of SOPs that were identified during WSP development for the WBDWSIP pilot projects.

ADB project team members meeting with members of the Women's Self-Help Group in Shreechandrapur, Gram Panchayat Barsal, District Bankura, West Bengal.

Documenting the operation, maintenance, and monitoring procedures is important because it

- helps build confidence that operators and the backup team know what to do and when,
- supports the consistent and effective performance of tasks,
- captures knowledge and experience that may otherwise be lost when community members have moved elsewhere,
- helps reinforce the importance of the role of the community in the water supply system,
- helps in the training and competence development of new community operators, and
- forms a basis for continuous improvement.

Source: World Health Organization. 2012. *Water Safety Planning for Small Community Water Supplies: Step-by-Step Risk Management Guidance for Drinking-Water Supplies in Small Communities*. Geneva.

Table 8: Some Standard Operating Procedures for the Pilot Projects

Implementation Phase	Process Step	Type of Document
Design	Water source	Water quality assessment guidance document
Construction	General	SOP for regular surveillance
Construction	General	SOP for third-party checking of soil testing reports
Construction	General	Quality control/Quality assurance plan
Construction	Transport	SOP for the introduction of new pipeline
Construction	Storage	SOP for water tightness testing of service reservoirs
Operation	Water source	SOP for periodic raw water pumping and storage in salinity-exposed water intake sites
Operation	Transport	SOP for cross-connections and for detection of unauthorized connections
Operation	Transport	SOP for new connections
Operation	Transport	SOP for valve maintenance
Operation	Transport	SOP for regular network inspection
Operation	Transport	SOP for pipeline repair
Operation	Transport	SOP for leakage control (procedure for reporting visible and invisible leaks)

continued on next page

Table 8 *continued*

Implementation Phase	Process Step	Type of Document
Operation	Transport	Water meter maintenance plan
Operation	Transport	SOP for the operation and maintenance of pumping machinery
Operation	Storage	SOP for the operation of service reservoirs (level control, cleaning and disinfection, routine inspection)
Operation	Treatment	SOP for the calibration and maintenance of WTP online monitoring equipment
Operation	Treatment	SOP for the start-up, operation, and shutdown of all WTP components
Operation	Treatment	SOP for chemical dosing
Operation	Treatment	SOP for WTP operation and maintenance (coagulation and flocculation, sedimentation, filtration, disinfection, etc.)
Operation	Treatment	SOP for the storage and handling of chlorine cylinders
Operation	Treatment	SOP for chemical quality assurance
Operation	Treatment	SOP for sludge handling
Operation	Treatment	WTP operation and maintenance plan
Operation	General	Water quality monitoring plan
Operation	General	SOP for the inspection and testing of electromechanical instruments
Operation	General	SOPs for energy audit and energy conservation
Operation	General	SOP for internal and external communication
Operation	General	SOPs for water meters, telemetry, and supervisory control and data acquisition (installation, operation and maintenance, testing and calibration)
Operation	General	SOP for internal and third-party auditing of operation and maintenance activities and key performance indicators
Operation	General	Asset management plan
Operation	General	Geographic information system management system
Operation	General	Security policy guidance document
Operation	General	Staff and contractor training programs
Operation	General	Personal safety and hygiene guidance document
Operation	General	Emergency plan

SOP = standard operating procedure, WTP = water treatment plant.

Source: Prepared by the consulting team, 2019.

Emergency Procedures

Emergency plans should be developed to address any unforeseen event or emergency. The water safety planning team should anticipate these conditions and establish procedures for dealing with them when they occur.

The most important emergency procedures should be readily accessible and should be part of the employee orientation process.

Through SWM, such emergency plans could be supported by prior forecasts and early warnings of such hazardous events, e.g., predictive maintenance (asset health monitoring with the help of sensors), raw water availability forecasts (drought forecasts), and water source contamination forecasts.

A list of potential emergency situations should be prepared, and procedures developed. Among these are the following:

- distribution system problems (equipment failure, treatment failure);
- power outages;
- loss of supply;
- contamination of supply; and
- vandalism and sabotage.

Tanushree Jana with her daughter, Riya Jana, at their house in Kalicharanpur Village, Purba Medinipur, West Bengal.

Some key areas to be included in emergency management procedures are as follows:

- actions to be taken in response to potential public health risks, including increased monitoring and inspection, and the issuance of boil-water or restricted-water-use advisories;
- an emergency notification list identifying those who should be contacted (by priority) and stating their names, addresses, and regular and alternative phone numbers, as well as the reason for the notification and the type of information required;
- roles and responsibilities of all stakeholders involved; and
- a description of alternative power and power sources, detailing access to each source and its operation.

Supporting Programs

Supporting programs are activities that, while not directly influencing water quality (like operating activities), are important in ensuring water safety.

Essential activities for water safety management include, but are not limited to (i) training programs; (ii) internal and external communication strategies (e.g., public awareness campaigns, involvement of public and institutional stakeholders); (iii) research programs; and (iv) tools for managing workers' activities, such as quality assurance systems and quality management systems (e.g., internal quality audits and annual external audits to revisit and review the water supply system).

Various guides and procedures are expected to be developed. These should include the following:

- emergency response plan,
- environmental management plan,
- health and safety plan,
- plan for managing stores and inventory,
- plan for managing customer and community complaints, and
- human resource training plan.

During water safety plan development, it became clear that the successful implementation of the plan required the establishment of more support systems, namely:

- **Training for personnel involved in system activities and the possible establishment of a training and resource center, particularly for smart water management (SWM).** Internal and external training should be provided to the staff, system operators, and contractor and/or agency

continued on next page

continued

workers so that they are made aware of quality issues, the importance of good and safe operation and maintenance practices, and SWM; and that they understand the importance of hygiene.

- **Community-led capacity-building programs** for *gram panchayat* (village council) members and *Pradhan* (head), village water and sanitation committee, and women's groups, as well as folk singers and others, for community mobilization and awareness raising. Efforts must be taken to minimize unaccounted-for water. Villagers must be motivated to use water in a way that eliminates or reduces water waste at the consumer's end.
- **Communication strategies (internal and external).** There should be clear lines of communication within the organization and between the organization and its external stakeholders. External communication channels should be agreed on with stakeholders and documented. Such mechanisms are important not only during emergency events but also in ensuring that the many stakeholders involved are informed in a timely way. There should be a procedure for disseminating information to company staff, as employees are ambassadors for their organization. Keeping them informed will minimize the spread of misinformation and enhance understanding of water safety planning procedures and protocols. Water production and treatment and water supply up to the reservoirs are the responsibility of the Public Health Engineering Department (PHED), while the local distribution of water from overhead reservoirs, household connections, revenue collection, and system maintenance at the local level are functions of local bodies. To ensure equitable distribution of water among the consumers in various villages, proper communication networks at various levels, including the field level, need to be established. A key aspect of communication is public information. Public awareness campaigns and the presentation of system evidence and results to the public and institutional stakeholders should be part of the communication strategy.
- **Information management.** Information and data should be focused on needs; they must be relevant and accurate, and in a form that is both accessible and easily interpreted. In the context of water safety planning, several organizations may require similar data. Working together may provide an opportunity for more efficient data collection, which will benefit all stakeholders involved in the planning.
- **Record-keeping and documentation.** These are records of regular monitoring of steps in the water safety planning process, reports on actions taken to correct deviations from critical limits, incident response reports, and other information relevant to water safety planning.
- **Smart water management**. The PHED will be combining real-time systems for monitoring and modeling raw water availability for the control of intake pumping and use of buffer reservoir storage.

Bula Ghosh at her home in South Adampur Village, Haroa, District 24 North Parganas, West Bengal.

4.4.2 Review of the Water Safety Plan

A water safety plan (WSP) is not meant to be a static operational document. As factors affecting the water supply system change, the plan will have to be revisited to ensure that it continues to identify and respond to emerging risks and hazards.

Periodically, the water safety planning team should meet to review and update the plan to make certain that it is being implemented and remains effective, and to understand and maximize the benefits arising from its preparation and implementation.

During the review, it may be helpful to do the following:

- Review and include any new activities or changes in the water supply chain (catchment area, abstraction, treatment, storage, distribution, and point of use), as applicable. This task includes reviewing and updating the water supply description and the system map or schematics as needed.
- Incorporate new hazardous events and hazards and associated risks in the WSP, and update previously identified risks with additional or new information. A hazardous event not found in the plan should be added to the plan, a full risk assessment made, and recommended actions included. The WSP should then be updated to incorporate any changes made after one of these events. The following questions can guide this updating process:
  - Was this hazardous event or hazard already in the plan?
  - If it was, what mitigation measures were used to deal with it?

  - Were those measures successful or not? Did they achieve the desired results?
  - Was the risk score correct? (If not, the risk score should be reviewed.)
- Require the responsible stakeholders to manage the actions and targets in the improvement plan. When these actions are completed, the risks identified in the WSP should be changed to reflect the improvements.
- Review roles and responsibilities and SOPs:
  - Have the roles and responsibilities of management or staff changed since the last review?
  - Have there been personnel changes since the last review?
  - Have there been any changes in system operation, maintenance, inspection, and monitoring processes and procedures?

Reviews can be further supported by data from smart water dashboards displaying measured critical parameters throughout the water supply system. The WSP should be reviewed fully at a frequency agreed on by the team members. This frequency can be anywhere between 1 and 3 years and should be agreed on following the completion of the first draft of the plan.
In addition to periodic programmed reviews, the WSP should also be assessed under the following special circumstances:

- after an incident, emergency, or near accident;
- after improvements or significant changes in the system; and
- after an audit or evaluation, to incorporate its findings and recommendations in the plan.

Following an emergency, an incident, or a near miss, the team should consider answering the following questions during the review:

- What was the cause of the problem?
- How was the problem first identified or recognized?
- What essential actions were required, and were these carried out?
- What communication problems arose, and how were these addressed?
- What were the immediate and the longer-term consequences?
- How well did the emergency procedures work?
- Did these hazardous events highlight weaknesses in the WSP, and how can the water safety planning team (or the local government) prevent a recurrence of the problem that caused the emergency?
- Has the WSP been updated to reflect the lessons learned to avoid having to deal with a similar problem in the future?

Students of Safique Ahmed Girls High School in Haroa, 24 North Parganas, West Bengal, filling their water bottles from the school's hand pump.

5

TOWARD AN INTEGRATED WATER AND SANITATION SAFETY PLANNING

5.1 Need for Integrated Planning

Water supply cannot be considered in isolation, especially at the household level in the *gram panchayats*. It is part of a wider water cycle, which also includes sanitation. This section discusses the issues and process involved in water and sanitation safety planning, combining water safety planning and sanitation safety planning, and using a framework similar to that used in water safety planning, particularly for the *gram panchayats*. Sample water and sanitation safety plans (WSSPs) for the project *gram panchayats*, in each of the districts covered under the West Bengal Drinking Water Sector Improvement Project (WBDWSIP), and water and sanitation safety planning guidelines for the state of West Bengal are to be prepared in 2021.

Adequate sanitation and good hygiene are indispensable elements of a decent standard of living. However, sanitation in the villages covered by the pilot projects is in a sorry state. With household water consumption at 70 liters per capita per day, large-scale drainage problems are possible unless the villagers adopt kitchen gardens and soak pits to contain gray water. Vector-borne diseases could increase, cesspools could form, and roads within the villages could be damaged.

The Government of India created the Swachh Bharat Mission–Grameen under the Ministry of Drinking Water and Sanitation (MDWS) to improve the quality of life in the rural areas by promoting cleanliness and hygiene, and eliminating open defecation. To this end, the Swachh Bharat Mission is

- accelerating sanitation coverage in the rural areas;
- motivating communities and *Panchayati Raj* institutions, through awareness-raising and health education campaigns, to adopt sustainable sanitation practices and facilities;
- encouraging the use of cost-effective and appropriate technologies for ecologically safe and sustainable sanitation; and

Bandana Mandal makes lentil-based snacks and is learning to grow mushrooms in Duber Danga Village, District Bankura, West Bengal.

- developing community-managed sanitation systems where required, focusing on scientific solid and liquid waste management for overall cleanliness in the rural areas.

In a parallel move, the Government of West Bengal set up the Mission Nirmal Bangla under the Panchayat & Rural Development Department to speed up the improvement of rural sanitation.

The basic purpose of these efforts is public health protection, and sanitation safety planning is a step in this direction. It assists in

- systematically identifying and managing health risks along the sanitation chain,
- guiding investment on the basis of actual risks to promote health benefits, and
- assuring the safety of sanitation-related products.

There are clear benefits to integrating the water safety planning and sanitation safety planning approaches into water and sanitation safety planning. Many risks within the water supply system are related to sanitation failures (e.g., poor wastewater drainage and treatment, system leakage). As water and sanitation systems are often in close proximity to each other, unsafe management of sanitation facilities could adversely affect not only the environment but also health through the contamination of water bodies, and thus drinking water supplies. Besides, with climate change posing a threat to water quality, water recycling is becoming an increasingly important source of water.

Water safety planning and sanitation safety planning can be used to coordinate the participation of the many stakeholders involved, including health, environment, and agriculture authorities, to support the long-term safe management of resources and protect public health. A multistakeholder, multisector platform for identifying the weaknesses and strengths of the water supply and sanitation systems, determining the needed improvements, and monitoring the systems is the basis for water and sanitation safety planning.

5.2 Developing a Water and Sanitation Safety Plan

Practical guidance for the step-by-step development of a WSSP using a preparation–assessment–monitoring–management framework is given below. The listed tasks are merely suggestions; the list can be expanded and adapted to the local situation in consultation with the stakeholders, as will be done for the *gram panchayats* by the project nongovernment organizations (NGOs) starting in late 2020 (with expected completion by early 2021).

5.2.1 Preparation

Task 1: Form a water and sanitation safety planworking team; identify roles and responsibilities

Table 9 presents the suggested composition of a water and sanitation safety planning team for the project *gram panchayats* in West Bengal, including the organizations represented and the roles and responsibilities of team members.

A stakeholder map identifying the relevant stakeholders in water supply and sanitation—who is responsible for what, and the level of influence—should be prepared.

5.2.2 Assessment

Task 2: Describe the water supply and sanitation systems in place

To identify risks and potential hazards and related diseases, and plan improvements, the mechanisms of the water supply and sanitation systems in place must be clearly understood.

Table 9: Possible Composition of Water and Sanitation Safety Planning Team for *Gram Panchayats*

Organization Represented	Position in the Organization	Roles and Responsibilities in WSSP Project	Participation Level
P&RDD	District coordinator	Coordination during implementation and monitoring of the entire WSSP	Water and Sanitation Safety Planning Task Force
P&RDD	Field supervisor	Risk assessment of drinking water sources and/or sanitation facilities	Water and Sanitation Safety Planning Task Force
P&RDD	Joint BDO	Coordination during implementation and monitoring of the entire WSSP	Water and Sanitation Safety Planning Task Force
P&RDD	Assistant coordinator, Sanitation	Coordination during implementation and monitoring of the entire WSSP	Water and Sanitation Safety Planning Task Force
P&RDD	Executive assistant	Coordination during implementation and monitoring of the entire WSSP	Core group
P&RDD	*Nirman Sahayak* (construction supervisor)	Coordination during implementation and monitoring of the entire WSSP Provision of technical assistance in infrastructure development to the *gram panchayat* (village council)	Core group
PHED	Executive engineer, Project implementation unit	District team leader and overall WSSP coordinator, with the following responsibilities: - assembling the required individuals and stakeholders for the WSSP coordination and maintenance teams; - providing appropriate training to the WSSP coordination and maintenance teams in the WSSP concept and implementation; - calling regular meetings of the water and sanitation safety planning team; - taking responsibility for all water and sanitation safety planning work, from plan development to implementation and sustainability; and - initiating plan revisits and possible revision and/ or improvement	Water and Sanitation Safety Planning Task Force
PHED	Assistant engineer	Bulk water supply WSSP assistant to the team leader, with the following responsibilities: - gathering data, - formulating the WSSP (hazard and risk assessments), - planning and monitoring improvements, - communicating regularly with district and block administration and the *gram panchayats*/VWSC,	Water and Sanitation Safety Planning Task Force

continued on next page

Table 9 *continued*

Organization Represented	Position in the Organization	Roles and Responsibilities in WSSP Project	Participation Level
		- developing and implementing site surveys and work plans for construction work, and - coordinating with line departments and grievance redress authorities	
PHED	Consultant, Water and Sanitation Support Organization	IEC, awareness generation, and public relations	Water and Sanitation Safety Planning Task Force
PHED	Environment specialist	Coordination of work done as part of the environmental impact assessments (including the health impact assessment) with the work related to water safety plan preparation	Water and Sanitation Safety Planning Task Force
PHED	Junior engineer	Regular surveillance of ongoing project	Water and Sanitation Safety Planning Task Force
PHED	Laboratory analyst	Water quality testing	Water and Sanitation Safety Planning Task Force
PHED	Social safeguard specialist	Coordination of the work done as part of the social impact assessments (including the health impact assessment) with the work related to water safety plan preparation	Extended team
PHED	District consultant, Water Quality	Water quality supervision and risk assessment	Support team
PHED	Lab coordinator, Water Quality	Water quality monitoring	Support team
Panchayat samiti (block-level council)	*Sahakari* (cooperative)	Coordination at different levels	Support team
Panchayat samiti	BDO (Gangajalghati)	Overall development works in block	Support team
Gram panchayat	*Anganwadi karmee* (child care center worker)		Core group
Gram panchayat	Community and/or religious leader	Awareness generation, mobilization, and message dissemination	Core group
Gram panchayat	*Gram panchayat pradhan* (village head)	The *gram panchayat* owns and/or manages the water supply scheme for the community and has the following responsibilities: - conducting liaison between the *Gram Sabha* (village meeting) and various programs; - approving the annual budget and user charges after discussion in the *Gram Sabha*; - approving memorandums of understanding and/or contracts with operators;	Core group

continued on next page

Table 9 *continued*

Organization Represented	Position in the Organization	Roles and Responsibilities in WSSP Project	Participation Level
		- hiring trained mechanics for preventive maintenance of hand pumps and piped water supply; - conducting liaison with the block-level WSSP team; - coordinating with the VWSC; - conducting periodic meetings, VHSNC strengthening, and follow-up meetings; - facilitating the development of social and village resource mapping and conducting sanitary surveys and O&M monitoring; - conducting, coordinating, and participating in capacity development training; and - engaging in community awareness and local coordination	
Gram panchayat	*Gram Sabha* member	Responsibilities are as follows: - facilitating resource mapping, - participating in sanitary surveys, - taking the lead in the installation of toilet and drinking water sources, and - participating actively in the O&M process	Core group
Gram panchayat	Local public health officer or local WASH expert (ASHA)	Responsibilities are as follows: - participating in resource mapping, - collecting data regarding water quality and sanitation survey of toilet and drinking water sources, and - disseminating information to raise awareness	Core group
Gram panchayat	Local public health officer or local WASH expert (AWW)	Responsibilities are as follows: - conducting resource mapping, door-to-door visits; - raising awareness; - conducting IPC activities; and - raising awareness (among expecting and new mothers, adolescents, etc.)	Core group
Gram panchayat	Member representing scheduled castes or scheduled tribes or poorer villagers	Participation in group meetings and participatory rural appraisal activities	Core group
Gram panchayat	MLA representative		Core group
Gram panchayat	SHG member	Motivation, participation in group meetings and IPC program activities Involvement of women in water safety activities	Core group
Gram panchayat	VWSC member	Responsibilities are as follows: - meeting regularly with VHSNC, ASHA, SHG, AWW, teachers, and other team members; - facilitating water quality surveys and sanitary surveys of all drinking water sources;	Core group

continued on next page

Table 9 *continued*

Organization Represented	Position in the Organization	Roles and Responsibilities in WSSP Project	Participation Level
		- conducting solid and liquid waste management planning; - conducting O&M activities; and - collecting and updating periodic reports and submitting them to the *gram panchayat*	
Gram panchayat	Volunteer water facilitator	Water sample collection, sanitary surveillance, and risk assessment	Core group
Gram panchayat	Youth representative		Core group
Department of Health & Family Welfare	BMOH	Health risk assessment	Support team
Department of Health & Family Welfare	Local public health officer or local WASH expert (ASHA)	Awareness generation related to water and sanitation issues	Support team
School Education Department	Schoolteacher	Awareness generation related to water and sanitation issues: safe drinking water, safe sanitation practices, and behavioral changes	Core group
SHG	Club or *Mahila Samity* (Women's Organization) member	Motivation, participation in group meetings and IPC program activities Involvement of women in water safety activities	Core group
Women and child development (ICDS)	Community development project officer		Core group
Civil society organization	Head and/or member		Extended team
Community Health Care Management Initiative	District coordinator	Monitoring of the WASH program (especially relating to community health)	Extended team
UNICEF	UNICEF representative	Coordination at different levels (sanitation, water quality)	Extended team (for consultation only)
Nongovernment organization engaged for the project	Team leader	Engaged to support *gram panchayats* and the WBDWSIP	Core group and in charge of developing WSSP for the *gram panchayats*

ASHA = Accredited Social Health Activist Program; AWW = *anganwadi* worker; BDO = block development officer; BMOH = block medical officer of health; ICDS = Integrated Child Development Services Program; IEC = information, education, and communication; IPC = infection prevention and control; MLA = Member of the Legislative Assembly; O&M = operation and maintenance; P&RDD = Panchayat & Rural Development Department; PHED = Public Health Engineering Department; SHG = self-help group; UNICEF = United Nations Children's Fund; VHSNC = village health, sanitation, and nutrition committee; VWSC = village water and sanitation committee; WASH = water, sanitation, and hygiene; WBDWSIP = West Bengal Drinking Water Sector Improvement Project; WSSP = water and sanitation safety plan.

Source: Prepared by the consulting team, 2019.

The Panchayat & Rural Development Department (P&RDD) is the nodal agency for the planning and implementation of the sanitation program for both solid and liquid waste in rural West Bengal, with offices at the state, district, and community levels. The *gram panchayats* (village councils), on the other hand, are responsible for managing water and sanitation services.

Therefore, while the P&RDD will be the overarching body providing guidance on the development of a water and sanitation safety plan, the *gram panchayats* will be the nodal agencies for plan implementation in rural West Bengal. They will need to demonstrate their commitment, with the help of the block and district councils, the P&RDD, and other governing bodies, by allocating the necessary resources (financial, staff, information, etc.) for plan implementation. Nongovernment organizations engaged in the project will assist the project *gram panchayats* in drafting the planning guidelines, to be rolled out for the state by the P&RDD. The nongovernment organizations will also conduct a strengths, weaknesses, opportunities, and threats analysis of the sanitation program at least at the block and village levels, and wide stakeholder consultation at all three levels of governance (including the district level), to determine the capacity and training needs for the proper implementation of the water and sanitation safety plan.

Members of the Women's Self-Help Group in Kalicharanpur Village, Purba Medinipur, West Bengal.

A detailed description of the whole system, from the water catchment area to abstraction, water treatment, water storage, water distribution and transport, and until the consumers' tap or storage place, is needed. There must also be a description of the sanitation system—mapping of the type of sanitation systems implemented, wastewater collection and treatment, and discharge or reuse. A map of the system, if there is one, would be very useful. If no such map is available, then it should be drawn up. The system map should clearly establish the path of different waste fractions through the system and the composition of the waste generated.

Available documents about the system, secondary data from government bodies, observations, and interviews with stakeholders should be collected and analyzed. Resource mapping, surveys, water quality analysis, and epidemiological investigation should be carried out, if not enough information is available. This data collection and analysis is very important since it will also serve as a baseline for assessing the impact of the implementation of the water and sanitation safety plan (WSSP).

A review of regulations and laws will also give an insight into the requirements of the related supply and sanitation system.

Task 3: Conduct risk assessment, identify exposure groups and routes, and agree on improvement actions

The risk assessment is aimed at identifying groups of people that may be exposed to particular hazardous events and hazards.

The exposure groups include those who could be harmed by their roles in working with waste, drinking water, or eating or handling contaminated crops, or simply by exposure to pollution during recreational activities. Among these exposure groups are utility workers, waste handlers, farmers, users, consumers, and the public in general. Exposure routes include ingestion of contaminated water or food; vector-borne transmission (with flies or mosquitoes); inhalation of aerosols and particles; and dermal contact with overflowing or leaking contents, contaminated stormwater drains, wastewater, sludge, slurry or manure, and contaminated groundwater or surface water.

The results of the overall assessment of the system, and the identified weaknesses and risks, should be documented and then shared and discussed with all stakeholders. The exposure groups can then be added to the system map.

Task 4: Develop and agree on an improvement action plan

An action plan for minimizing the risks related to the water supply and sanitation systems in place should be developed. Realistic targets should be set for improving the systems, controlling the risks and hazards, and minimizing health risks in particular. For several reasons, needed and wished-for improvements often cannot be realized immediately, but step-by-step improvements can be planned. More expertise or training may be requested.

5.2.3 Monitoring

Task 5: Identify and document the needed implementation resources

The financial and human resources needed to implement the planned actions and the time frame should be identified and documented.

5.2.4 Management

Task 6: Develop and implement management procedures

The management procedures include written instructions that describe
(i) the steps or actions to be taken during normal operating conditions, and
(ii) corrective actions or emergency plans to be carried out when operational monitoring parameters reach operational or critical limits or exceed those limits. Such procedures reinforce the WSSP indirectly and are necessary for the proper operation of the control measures.

A key aspect is communication. The community and relevant stakeholders should be consulted and involved in the planning and implementation of the WSSP with all its features. The plan should be transparent and informative, it should be made public, and its eventual consequences should be made known.

5.3 Water and Sanitation Safety Planning Challenges

Given the demographic and socioeconomic characteristics, and the institutional and infrastructure capacity of the rural water supply and sanitation service sector, developing and implementing a comprehensive WSSP is a major challenge. Key constraints and concerns include the following:

- **Limited data availability.** Basic data concerning catchment and source water quality and also various components of water transmission, distribution, storage, and household handling in the rural areas are not easy to obtain. Assessment of risks from chemical and microbial contamination of the system from catchment to the consumers' end becomes difficult in the absence of regular sanitary inspection data and information.

Selina Akhtar, an 8th standard student in the in Brij Mohan Tewary Girls High School in Nandigram, Purba Medinipur, West Bengal.

- **Poor sanitation.** Poor status of rural sanitation, including facilities for human excreta disposal, drainage, and solid waste management, and also open defecation in many areas, increase the risks of fecal contamination of the system in the rural areas.
- **Inadequate knowledge of piped water supply systems.** Because of lack of cultural expertise in community water supply management, poor record keeping, or lack of post-construction documentation in most village-level organizations in the rural areas, much of the information regarding the piped water supply system and its management may not be available. Moreover, in most cases, the community is not used to having such a large quantity of treated water at the household level. To avoid waste and environmental pollution, other facilities like bathing places and proper drainage systems may have to be made available within the premises.
- **Low availability of equipment and skilled manpower.** Availability of resources as well as infrastructure such as laboratories, skilled manpower, and technical expertise is a challenge in rural areas. The WSSP must also address these problems, including capacity-building and resource mobilization requirements.

Members of the Women's Self-Help Group during a meeting with ADB team in South Adampur Village, Block Haroa, Gram Panchayat Haroa, District 24 North Parganas, West Bengal.

6

HYGIENE AND BEHAVIORAL CHANGE CAMPAIGNS AT THE *GRAM PANCHAYATS*

Though the water safety planning guidelines presented here are targeted mainly at bulk supply service providers and operators of rural water schemes, without adequate emphasis, and institutionalization of practices for a continuous water, sanitation, and hygiene (WASH) improvement and behavioral change management at the *gram panchayat* levels, efforts by bulk supply service providers alone will not be sufficient to achieve the desired public health improvement. The West Bengal Drinking Water Sector Improvement (WBDWSIP) has engaged three nongovernment organizations (NGOs) to support the government and the *gram panchayats* in this regard and also in the implementation of the smart water management (SWM) of water supply services.

The NGOs are assisting the project *gram panchayats* in community mobilization, consultation, and participation; the design and implementation of awareness generation, WASH, and behavior change communication (BCC) campaigns; the identification and inclusion of vulnerable groups in project benefits; and the implementation of gender equality and social inclusion action plans.

The project NGOs have prepared detailed work plans for stakeholder consultation and participation, and for WASH and BCC programs. **The findings from the implementation of these WASH and BCC plans in project *gram panchayats* will provide good lessons that can be incorporated in the WSSP guidelines expected to be prepared for the *gram panchayats* in early 2021 once the pilot-testing of water scheme management for a few project *gram panchayats* is completed.** The findings of the baseline impact evaluation survey done in 2019 reveal that a large proportion of people in rural West Bengal wash hands with water alone and do not use soap. The WASH and BCC programs are designed to address specific issues pertinent to the present context and needs of rural West Bengal, given the survey findings as well as the current coronavirus disease (COVID-19) pandemic.

The WASH and BCC plans of the project NGOs are focused on hand washing, personal hygiene, the use of clean water for hand washing and bathing, and the importance and meaning of safe sanitation.

In response to the global COVID-19 pandemic, the WASH and BCC plans under the WBDWSIP have been updated to include information and awareness-raising materials on the spread of epidemics and pandemics, the importance of early identification and quarantine, practical examples related to required precautions (including social distancing) to be taken to avoid community transmission, other measures required for the protection of communities during pandemics, steps to be taken when someone in the community tests positive, and other matters. Protecting rural settlements from the spread and impact of epidemics and pandemics like COVID-19 is an integral part of the WASH–BCC campaign.

The NGOs have done vulnerability mapping and prepared guidelines to support local action during such disease outbreaks.

The community can help the government to (i) identify volunteers and caregivers who can deliver services to the elderly, persons with disabilities, children, transgender people, and other vulnerable sectors; (ii) create awareness on prevention, social distancing, and isolation; (iii) combat stigma; and (iv) provide food and relief to daily wage workers, and to poor and vulnerable families, e.g., through community kitchens. These elements are being factored into the design of the WASH and BCC campaigns under the WBDWSIP. WASH plans prepared by the project NGOs deal in particular with the following: (i) health and safety, including a suitable response to challenges posed by epidemics and pandemics like COVID-19; (ii) water security and safety; (iii) regular disinfection of water sources and water storage; (iv) drinking water advisory communication tools; (v) sanitation and wastewater management; (vi) hand hygiene; (vii) community WASH; (viii) emergency and disaster preparedness; and (ix) preparation of training modules for capacity building. The WASH and BCC approach adopted for the WBDWSIP, given the project's overall objectives, involves the following:

- capacity building for WASH professionals;
- assessment of the status of WASH infrastructure in rural settings;
- supportive supervision;
- rigorous training for BCC through information, education, and communication and BCC modules and materials;
- a theme-based information, education, and communication campaign for social behavior change;
- the setting of realistic targets and recommendations for sustaining the good practices;
- WASH improvements in health facilities, among doctors, patients, and visitors;
- WASH improvements in schools and Integrated Child Development Services centers;
- access to WASH services in public places, including orientation on maintaining the cleanliness of facilities used, out of concern for others;
- sanitation and hygiene improvements, including hand hygiene and food hygiene at home, performed routinely while giving care to the children and the sick;

- communication and the establishment of social norms to end open defecation; and
- community preparedness for, and response to, epidemics and pandemics like COVID-19.

Table 10 lists the activities and expected outputs of the WASH and BCC plans of the *gram panchayats* under the WBDWSIP.

Table 10: Hygiene and Behavioral Change Outputs at the *Gram Panchayats*

Activity	Target Group	Expected Output
WASH baseline survey to understand the present situation in target areas	Households, schools, communities, institutions, etc.	Better understanding of the current status of WASH issues
District-, block-, panchayat-level sensitization meeting on WASH, and sharing of baseline report findings	District, block, and *panchayat* officials	Administrative officials sensitized to WASH issues and able to address community-level concerns
Probabilistic risk assessment (problem tree, tail method) of contagious diseases and the adverse effects of unsafe drinking water and sanitation	Community members, students	Target group able to identify the adverse effects of unsafe drinking water and sanitation, and aware of the importance of WASH issues
Public service advertisements (wall writing, posters, billboards, banners) on WASH	Households, schools, communities	Mass awareness
WASH workshop for local stakeholders in the WBDWSIP *gram panchayats* to discuss community responsibilities and actions related to infectious diseases like COVID-19, and other matters	ASHA (frontline health workers), *anganwadi* (child care center) workers, self-help group members, community leaders, village water and sanitation committee members, school teachers	Local stakeholders sensitized and able to interpret and respond to issues at the grassroots level
WASH awareness and sensitization program in schools	School students	School students sensitized and expected to convey the message to their parents
Formation of a WASH committee in every school	School students	Schools able to meet the basic criteria for dealing with WASH and sanitation- and hygiene-related issues
Provision of two types of garbage dustbin (for biodegradable and nonbiodegradable garbage) in every school	Schools	Students able to understand the differences between biodegradable and nonbiodegradable waste, and to convey the message to other people
Extensive IEC and BCC campaigns on the importance of the quality of drinking water, and proper sanitation and hygiene, particularly in the light of the COVID-19 outbreak and its control	Communities, schools	Mass awareness
Endline survey to understand the impact of the project	Households, schools, communities	Project evaluation against project objectives

ASHA = Accredited Social Health Activist Program; BCC = behavior change communication; COVID-19 = coronavirus disease; IEC = information, education, and communication; WASH = water, sanitation, and hygiene; WBDWSIP = West Bengal Drinking Water Sector Improvement Project.

Source: Prepared by the consulting team, 2019.

Students of Safique Ahmed Girls High School in Haroa, 24 North Parganas, West Bengal, filling their water bottles from the school's taps.

ACHIEVEMENTS AND LESSONS LEARNED

The biggest advantage of adopting water safety planning over end-product testing is ensuring that the processes involved in delivering safe drinking water are under full control at all times. In addition, water safety planning helps to increase the consistency with which safe water can be supplied and suggests contingency plans for responding to system failures or unforeseeable hazardous events. It also provides information about asset condition and management, and associated current and future investment requirements, and guides the implementation of both engineering and management system controls to prevent failures.

This project, the first experience with applying the water safety planning approach to a nonoperating bulk water supply system, has offered some lessons:

- The bulk water safety planning pilot projects in Bankura and North 24 Parganas districts presented opportunities to demonstrate that water safety planning principles are applicable and relevant to the design and construction phases of water service delivery implementation.
- A poorly designed or defectively constructed system will make it very difficult to provide safe water in the long term. By assessing what could go wrong and identifying appropriate control measures that can be implemented, it is possible to save money and improve the targeting of resources.
- Improvement actions that can be quickly implemented should not wait for the completion of the water safety plan (WSP) but should be implemented as soon as possible. That would materialize the concept of risk assessment and management, and enhance understanding of what water safety planning means in practice.
- During the risk assessment, an important hazardous event identified in the design phase was the lack of a detailed assessment of raw water quality to help in defining water treatment design criteria. To minimize the impact, the team acted quickly and requested a more complete analysis of the water source to determine if the defined treatment would be sufficient or if changes were needed.
- The hands-on approach to the water safety planning training workshops proved very successful as it enabled the teams to be actively engaged throughout the process of developing a WSP.

- The workshops created significant opportunities to work across teams and reflect on where they stand today. Most importantly, the workshops provided orientation in better planning and decision making. The importance of evidence, information documentation, and well-defined processes in place was also made clearer.
- Full meetings with a large group of people who have different roles and hold different views are not easily arranged. To facilitate the implementation of the pilot projects, the water safety planning team members were divided into two groups: core and extended. The core team was responsible for developing and implementing the WSP, and the extended team was involved when necessary and will be part of the regular meetings that should take place once the system is operational.
- It is very difficult to develop a WSP without the detailed involvement of the relevant state and district authorities and local practitioners. Stakeholders must understand the benefits that they can derive from this process in the performance of their duties.
- To improve the functioning of both water supply and sanitation systems, the water safety planning team felt that the delivery mechanism should be under a single department, given the strong links between the two systems.
- The process of exploring hazardous events was complicated by the lack of time and reliable information or difficulty in reaching agreement among the team members.
- The application of the risk identification step is not always straightforward; it is important to include at least one risk specialist to facilitate the process. The participation of an external facilitator would allow stakeholders to

Sandhya Sardar works as a cook at the *anganwadi* community in Baltiya Village, District 24 North Parganas, West Bengal.

Chinmayee Maity, Sujata Maity, and Ujjainee Maity are residents of Nandan Nayak Bar Village, Purba Medinipur, West Bengal.

challenge the effectiveness of mitigations and ensure that the correct scores are recorded with their agreement.

- Over time, support for the WSP can be difficult to sustain because of the long duration of projects and discontinuities in staff roles. Stakeholders may lose interest in the plan, and changing incumbents and roles during project implementation will not help matters. To help reduce the risk of diminishing support for the WSP, suitable up-front agreements with counterpart agencies and stakeholders could be implemented to gain their commitment to the plan.
- Given the results of this first application, extending this experience to other districts is recommended. The valuable lessons learned can be applied in other districts, as the local water sector context, difficulties, and support needs are similar.
- It is recommended that a larger time frame be considered in future implementation to increase the reliability of the information provided and enable a more productive reflection on water safety planning practices.
- Adoption of water safety planning for the WBDWSIP will require the PHED to take a proactive approach. For the planning team to function properly with responsibility and authority, administrative orders may have to be issued.

Members of Women's Self Help Group in Baltiya Village, Dsitrict 24 North Parganas, West Bengal, India.

APPENDIXES

The following pages provide templates and examples based on the West Bengal Drinking Water Sector Improvement Project pilot projects in the districts of Bankura and North 24 Parganas:

Appendix 1 Water Safety Plan Objectives and Indicators

2 The Water Safety Planning Team

3 Stakeholder Analysis

4 Checklist of Information about the Water Supply System to Be Compiled

5 Environmental Legislation, Strategy, and Guidance Documents

6 Brief Description of the Bulk Water Supply Systems for Bankura and North 24 Parganas Districts

7 Hazardous Events and Hazard Risk Assessment

8 Improvement Plans

9 Operational Monitoring of Control Measures

Appendix 1: Water Safety Plan Objectives and Indicators

Table A1: Sample Water Safety Plan Objectives and Indicators

Objective/Outcome	Indicator	Formula
Improve availability, reliability, and accessibility	- % of population with access to improved source of water[a] - Average no. of liters of potable water available per person per day - % of 24/7 supply (for wettest season and for driest season)	- (Population with access to improved source of water/Total community population) x 100 - (Amount of water produced – unaccounted water [per day, in liters])/Actual total population depending on the system as its primary source - ([Hours per day of service x days per week of service]/168) x 100
Minimize cases of waterborne diseases	No. of cases of waterborne diseases[b]	No. of cases of diarrhea, dysentery, cholera, typhoid, infectious hepatitis A, and worm infestation
Improve water quality	- % compliance with microbiological parameters - % of water quality tests with FRC according to BIS drinking water quality standards	- (No. of analyses of fecal coliform in compliance [=0]/Total no. of samples) x 100, at OHSR or GLSR outlet - (Water quality tests with FRC according to BIS drinking water quality standards/Total number of water samples) x 100, at OHSR or GLSR outlet
Improve management and operational procedures	Standard operating procedures[c]	**Score = 100:** There is a written plan/procedure that is clear, thorough, accurate, documented, and fully implemented in practice. **Score = 50:** There is a plan/procedure, but it is not documented, not clear or accurate, and/or not practiced. **Score = 0:** There is no plan/procedure.
Improve management and operational procedures	Availability of routine maintenance schedule and compliance with the schedule; availability of spare parts and technical support for component repairs beyond the capacity of operating personnel	**Score = 100:** Routine maintenance is done according to the maintenance schedule; relevant staff have been trained in the use of the O&M manuals; spare parts and well-equipped technical support are available, within 24 hours, for maintenance or repairs. **Score = 75:** Routine maintenance is done according to the maintenance schedule; relevant staff have been trained in the use of the O&M manuals; and spare parts and well-equipped technical support are available. However, it takes from 1 to 3 days to carry out maintenance or repairs. **Score = 50:** Routine maintenance is done according to the maintenance schedule, and spare parts and well-equipped technical support are available. However, relevant staff have not been trained in the use of the O&M manuals, and it takes from 1 to 3 days to carry out maintenance or repairs. **Score = 25:** Routine maintenance is randomly executed, there are no spare parts, and while well-equipped technical support is available, it takes from 1 to 3 days to carry out maintenance or repairs.

continued on next page

Table A1 *continued*

Objective/Outcome	Indicator	Formula
		Score = 0: No routine maintenance is carried out, and there are no spare parts or well-equipped technical support available to carry out repairs, or where available, it takes longer than 3 days to carry out repairs.
Improve management and operational procedures	Operator training programs (plans for training system operators or staff)[c]	**Score = 100:** Operators and staff have received initial and refresher training, as documented in manually updated human resource capacity-building records of employees or in computerized human resource records. **Score = 50:** Operators and staff members have received initial training. **Score = 0:** Operators and staff members have not received any training.
Achieve financial sustainability	Positive annual revenue and/or expenditure balance	**Score = 100:** Annual revenue is higher than annual expenditure. **Score = 0:** No records of financial data are available; there is no revenue generation; annual revenue is lower than annual expenditure.
Achieve financial sustainability	Total revenue as % of total operating cost[c]	(Annual revenue collected/Annual operating cost) x 100
Improve monitoring and surveillance	Operational monitoring plan[c]	**Score = 100:** There is a written plan/procedure that is clear, thorough, accurate, documented, and fully implemented in practice. **Score = 50:** There is a plan/procedure, but it is not documented, not clear or accurate, and/or not practiced. **Score = 0:** There is no plan/procedure.
Improve monitoring and surveillance	Yearly compliance water quality sampling and analysis performed by recognized institutions and paid for by each community	**Score = 100:** Water quality testing and analysis is done yearly by certified laboratories according to BIS standards. **Score = 50:** Water quality testing and analysis is done by certified laboratories, but not as often as required. **Score = 25:** Water quality testing and analysis is done, but less than once a year or not by certified laboratories. **Score = 0:** Water quality testing and analysis is not done at all.
Increase communication and collaboration	Existing evidence that water safety planning meetings are held and information is shared with community members	**Score = 100:** Meetings are documented. **Score = 50:** Meetings are not always documented. **Score = 0:** No meetings are held.

BIS = Bureau of Indian Standards, FRC = free residual chlorine, GLSR = ground-level service reservoir, O&M = operation and maintenance, OHSR = overhead service reservoir.

[a] The source should be located on the premises, available when needed, and free from fecal and priority chemical contamination.

[b] The information should come from a reliable existing source, e.g., a standard health monitoring institutional instrument already in place (such as a medical office in the *gram panchayat* or village council).

[c] Metric adapted from lessons learned from Asia and the Pacific (Kumpel et al. 2018. Measuring the Impacts of Water Safety Plans in the Asia-Pacific Region. *International Journal of Environmental Research and Public Health.* 15 (6). pp. 12–23) and data collection form for water safety plan impact assessment (WHO, pre-publication version, October 2017).

Source: Prepared by the consulting team, 2019.

Bula Ghosh taking water from a hand pump installed at her house in South Adampur, Haroa, District 24 North Pargana, West Bengal.

Appendix 2: The Water Safety Planning Team

Table A2: Possible Composition of Water Safety Planning Team[a]

Core/ Extended Group	Organization	Position	Roles and Responsibilities in Water Safety Planning Project
Core	Public Health Engineering Department (PHED)	Executive engineer Project Implementation Unit (PIU)	Team leader with the following responsibilities: - leading water safety plan (WSP) development and overall coordination; - assembling the required individuals and stakeholders for the WSP coordination and maintenance teams; - providing appropriate training, as required, in the WSP concept and implementation; - calling regular meetings of the water safety planning team; - taking responsibility for all water safety planning work, from WSP development to implementation and sustainability; - initiating WSP revision and/or improvement; - coordinating with external and internal stakeholders; and - having primary responsibility for water supply system management, water quality, and water safety risks
Core	PHED	Assistant engineer (PIU)	Assistant team leader with the following responsibilities: - working with the water safety planning coordinator, contributing to WSP development, and supporting implementation; - coordinating the risk assessment of the water supply system and initiating the required monitoring to evaluate water safety; - supporting WSP development and implementation (collecting data, reviewing the team's WSP output and identifying control measures to mitigate the potential hazards identified, supporting the review of the WSP, recording/documenting the WSP development process, and preparing minutes of the water safety planning team meetings); - maintaining regular communication with district/block administration and the *gram panchayat* (village-level governance) water and sanitation committee; - developing and implementing a construction survey and work plan for the site; - coordinating public consultations and information disclosure; and - coordinating with line departments and assisting in the project grievance redress process, including the resolution of issues related to social and environmental safeguards
Core	PHED	Junior engineer (PIU)	Responsible for the following: - supporting implementation and contributing to the WSP (describing the distinctive features of the water supply system, identifying hazardous events and improvement actions, conducting operational monitoring, implementing and verifying the operational procedures implemented); and - monitoring and supervising project activities (construction works, including safety and hygiene procedures)

continued on next page

Table A2 *continued*

Core/ Extended Group	Organization	Position	Roles and Responsibilities in Water Safety Planning Project
Core	PHED	District consultant (Water Quality)	Responsible for the following: - supervising water quality; - preparing water quality assessment guidelines (water sampling, frequency of monitoring and monitoring parameters, analytical methods); - supervising data compilation (by water custodian authorities), validation and statistical analysis; and - supporting implementation and input to the WSP
Core	Water and Sanitation Support Organization, PHED	Consultant	Responsible for the following: - conducting awareness-raising and public relations programs; - developing information, education, and communication materials; - providing training (in water quality, hygiene, water conservation); and - assisting in water quality surveillance
Core	PHED	Laboratory coordinator	Responsible for the following: - providing water quality monitoring and surveillance; and - providing water quality assessment guidelines (data compilation by water custodian authorities and validation, water sampling, monitoring frequency and parameters, statistical analysis, etc.)
Core	PHED	Design supervision and institutional support consultant	Responsible for the following: - preparing subproject appraisal reports, detailed engineering design and bid documents, procurement and construction management reports; - preparing bid specifications and bid evaluation criteria; - providing contract administration, construction supervision, and quality assurance and quality control services according to the provisions of the construction contracts; - ensuring compliance with safeguard requirements; and - coordinating with other government authorities
Core	PHED	Planning and design circle superintending engineer	Responsible for the following: - providing engineering design and preparing specifications (for conformity of materials during project design); - advising on procurement in accordance with the government's and the Asian Development Bank's procurement guidelines; - engaging in geographic information system (GIS) mapping; - supporting procurement activities; and - contributing to WSP development and support implementation (carrying out studies related to pressure/ flow and identifying critical points in the network (terminals, sensitive points, etc.)
Core	Bulk Water Supply System Operator	Project manager	Responsible for the following: - managing the entire bulk water supply system, - assuming overall responsibility for monitoring major activities in the water supply system, and - implementing and provide input to the WSP

continued on next page

Table A2 *continued*

Core/ Extended Group	Organization	Position	Roles and Responsibilities in Water Safety Planning Project
Core	Bulk Water Supply System operator	Water treatment plant (WTP) manager	Responsible for the following: - taking charge of operation and maintenance of the WTP; - supervising the day-to-day operation of the WTP, including the quality of the materials in the warehouse, and controlling incoming chemicals to ensure that products purchased are suitable for the intended purpose (water for human consumption); and - implementing and providing input to the WSP (data needed for operations, such as volume treated, chemicals used, and hours of operation)
Core	Bulk Water Supply System operator	Operations coordinator	Responsible for the following: - coordinating the day-to-day operation and maintenance of the WTP and the bulk water supply system (pumping station, service reservoirs, transmission mains); and - implementing and providing input to the WSP (identification of hazardous events, improvement actions, operational monitoring, implementation and verification of operational procedures)
Extended	Smart water management (SWM) consultant	Team leader	Responsible for providing guidance and initial support in establishing new SWM applications in the use of measurements, data analyses and data-driven decision-making, controls, and alerts
Extended	Department of Health & Family Welfare	Block medical officer of health	Responsible for the following: - supporting the health risk assessment and other health-related activities (health liaison, e.g., compliance monitoring, emergency response planning); - providing data and latest study findings on health-related matters relevant to the operation of the water supply systems; - enlightening the community about the importance of safe water and good hygiene practices; and - assisting in the awareness campaign
Extended	Panchayat & Rural Development Department	District coordinator	Responsible for coordinating the planning and implementation of the sanitation program for rural West Bengal (Swachh Bharat Mission/Mission Nirmal Bangla)
Extended	Water Pollution Control Board	Representative	Responsible for the following: - assisting in environmental monitoring and pollution control, and - assisting in securing environmental clearance
Extended	Disaster Management	Representative	Responsible for supporting risk assessment and mitigation actions
Extended	State Water Investigation Department	Representative	Responsible for assisting in water resource assessment and management

continued on next page

Table A2 *continued*

Core/ Extended Group	Organization	Position	Roles and Responsibilities in Water Safety Planning Project
Extended	West Bengal State Electricity Distribution Company Limited	Representative	Responsible for the following: - assisting in ensuring uninterrupted quality power supply; and - assisting in utility service shifts (provision of power cables, etc.)
Extended	Public Works Department (PWD–Roads)	Representative	Responsible for providing permission needed for alignment of water supply pipeline
Extended	*Zilla parishad* (District-level governance)	*Sabhadhipati* (Chair)	Responsible for attending to district-level development planning and monitoring (water supply, sanitation, roads, etc.)
Extended	*Panchayat samiti* (Block-level governance)	*Sabhapati* (Chair)	Responsible for attending to block-level development planning and monitoring (water supply, sanitation, roads, etc.)
Extended	*Panchayat samiti*	Block development officer	Responsible for providing local public administration services
Extended	Land and Land Reform Department	Block land and land reform officer	Responsible for resolving land issues
Extended	*Bazar samiti* (Market-level governance)	Representative	Responsible for resolving any inconvenience regarding pipe laying around the marketplace
Extended	*Zilla parishad*	Assistant engineer	Responsible for assisting in securing a no-objection certificate
Extended	*Zilla parishad*	*Karmadhakshaya* (Representative)	Responsible for sensitizing the line department and the elected representative to the project to ensure its smooth operation
Extended	*Gram panchayat*	*Pradhan* (Head)	Responsible for the following: - providing local coordination, - engaging in community awareness programs, - providing liaison between the *Gram Sabha* (village meeting) and various programs, - approving the annual budget and user charges after discussion in the *Gram Sabha*, - approving memorandums of understanding/contracts with operators, - coordinating with the block resource center, and - hiring trained mechanics for preventive maintenance of hand pumps and piped water supply
Extended	*Gram panchayat*	Member	Responsible for engaging in community awareness and local coordination programs

[a] Based on the WSP team tables developed for the Bankura and North 24 Parganas pilot projects under the West Bengal Drinking Water Sector Improvement Project.

Source: Prepared by the consulting team, 2019.

Kalpana Manna (with water pot and bucket) resident of Kalicharanpur Village, District Purba Medinipur. Villagers are largely using ground water for their needs.

Appendix 3: Stakeholder Analysis

Table A3: Sample Stakeholder Analysis

Stakeholder	Roles and Responsibilities	Why the Stakeholder Should Be Involved	Motivating Factors[a]	Constraining Factors[b]	How the Stakeholder Should Be Involved[c]
ADB WBDWSIP project implementation unit and project management unit	Implementing and monitoring the WBDWSIP	Monitor and review the activities of the project management consultant for smooth and timely project implementation	Drinking water security through 24/7 PWSSs in the areas covered, institutional strengthening, and stakeholder capacity building at all service delivery levels for sustainable O&M and public health improvement	None	Meetings among beneficiaries to raise awareness for WSP implementation in the district
ADB WBDWSIP project management consultant	Supporting WBDWSIP implementation	Coordinate and facilitate contractual obligations for successful project implementation, and overall project management to ensure sustainability of the PWSSs handed over	Drinking water security through 24/7 PWSSs in the areas covered, institutional strengthening, and stakeholder capacity building at all service delivery levels for sustainable O&M and public health improvement	None	Meetings among beneficiaries to raise awareness for WSP implementation in the district
Petroleum and Explosives Safety Organisation (formerly Department of Explosives)	Ensuring the safety of water treatment plant workers Inspecting the chlorination plant	Issue license/ NOC as a statutory organization	Safety in the workplace	Knowledge gap	Issuing of clearance for storage and handling of chlorine gas as disinfectant

continued on next page

Table A3 *continued*

Stakeholder	Roles and Responsibilities	Why the Stakeholder Should Be Involved	Motivating Factors[a]	Constraining Factors[b]	How the Stakeholder Should Be Involved[c]
Department of Health & Family Welfare	Providing public health services in the state	Oversee environmental health	Provision of public health services	Environmental degradation	Meetings among key stakeholders at the state level and work plan development for project implementation
Panchayat & Rural Development Department	Upholding the Constitution and framing of policy on the functioning of rural local self-government (*panchayats*), providing administrative support to the *panchayat* system, and implementing rural development programs. The associated local *zilla parishads* (district-level governance units), *panchayat samitis* (block-level units), and *gram panchayats* (village-level units) will also be stakeholders. The department is the nodal department for rural sanitation in West Bengal, responsible for the planning and implementation of the rural sanitation program, and controls the budget for that purpose. (The Urban Development & Municipal Affairs Department looks after urban sanitation and sewerage activities.)	Ensure environmental sanitation under different programs such as the Mahatma Gandhi National Rural Employment Generation Scheme, the National Vector-Borne Disease Control Programme, solid and liquid waste management, and the Mission Nirmal Bangla Local coordination for capacity building of various stakeholders (*Panchayati Raj* institutions)	Sanitation safety at *gram panchayat* level, and employment generation through 100 days of guaranteed wage employment	Lack of awareness of safe sanitation practices	State- and district-level coordination committee and/or team
District Water and Sanitation Mission	Responsible for the following: - formulating, managing, and monitoring progress of drinking water security projects and the Total Sanitation Campaign in rural areas;	Frame implementation and monitoring strategy at the district level	Implementation of different schemes in the field	None	Water safety planning general meetings at the district level

continued on next page

Table A3 *continued*

Stakeholder	Roles and Responsibilities	Why the Stakeholder Should Be Involved	Motivating Factors[a]	Constraining Factors[b]	How the Stakeholder Should Be Involved[c]
	- examining and approving schemes submitted by the block *panchayat/gram panchayat* and forwarding these to the SLSSC where necessary; - selecting and entering into agreements with agencies and nongovernment organizations for social mobilization, capacity development, communication, and project management and supervision; - sensitizing public representatives, officials, and the general public to the need for drinking water security; - developing a pool of district resource persons with the experience and expertise to assist the various water safety planning teams; - engaging institutions to provide training to stakeholders, and carrying out a communication campaign; - coordinating water and sanitation matters between district health, education, forestry, agriculture, rural development, and other representatives, as well as with national programs; and - interacting with the SWSM, the state government, and the national government				
Arsenic Task Force	Serving as advisory committee for reducing the arsenic problem in drinking water in West Bengal	Assist the government in framing policies to reduce the arsenic problem in quality-affected dwellings in West Bengal	Compliance with the mandate	No constraining factors (the implementation of decisions taken by this advisory task force depends solely on the government)	Meetings among key stakeholders at the state level and work plan development for project implementation

continued on next page

Table A3 *continued*

Stakeholder	Roles and Responsibilities	Why the Stakeholder Should Be Involved	Motivating Factors[a]	Constraining Factors[b]	How the Stakeholder Should Be Involved[c]
Fluoride Task Force	Serving as advisory committee for reducing the problem of excess fluoride in drinking water in West Bengal	Assist the government in framing policies to reduce the fluoride problem in quality-affected dwellings in West Bengal	Compliance with the mandate	No constraining factors (the implementation of decisions taken by this advisory task force depends solely on the government)	Meetings among key stakeholders at the state level and work plan development for project implementation
Forest Department	Promoting and preserving forestation and preventing deforestation, and protecting wildlife	Secure permission for tree felling	Granting of NOC		District water safety planning general meetings
Gram panchayat and VWSC	Responsible for the following: - overseeing development work at the village level; - planning, designing, and implementing all in-village drinking water and sanitation activities; - providing facts and figures to the *gram panchayat* for a review of water and sanitation issues; - providing input to the WSP; - ensuring community participation and decision making in all phases of in-village scheme activities; - organizing community contributions toward capital costs, both in cash and in kind (land, labor, or materials); - opening and managing a bank account for community cash contributions, O&M funds, and project management funds; - commissioning and taking over completed in-village water supply and sanitation works following a joint inspection with line department staff;	Ensure community participation in planning, implementation, and maintenance of drinking water supply system, for smooth and timely project implementation	Adequate water supply and sanitation within the project area Training/ Capacity building/ Sensitization of VWSC members	Limited time for active involvement of VWSC members from planning to implementation of the schemes	Water safety planning meetings at the block and *gram panchayat* levels *Gram Sabha* (village meeting where all development issues are discussed, and report is submitted to the *gram panchayat pradhan* or the block district officer)

continued on next page

Table A3 *continued*

Stakeholder	Roles and Responsibilities	Why the Stakeholder Should Be Involved	Motivating Factors[a]	Constraining Factors[b]	How the Stakeholder Should Be Involved[c]
	- collecting funds through a tariff, charges, and deposit system for O&M of water supply and sanitation works, for the proper management and financing of O&M on a sustainable basis; and - empowering women for the day-to-day O&M of the scheme				
Irrigation and Waterways Department	Responsible for providing irrigation facilities, offering reasonable flood protection, easing drainage congestion, preventing erosion, and maintaining internal navigation channels and natural waterways in the state The department has implemented several major and medium irrigation projects, a number of embankment schemes, town protection schemes, drainage schemes, anti-riverbank erosion schemes, and anti-sea erosion schemes		Granting of water withdrawal permit and NOC for land use		State-level meetings among key stakeholders and in work plan development for project implementation
Panchayat samiti	Overseeing development work at the block level	Ensure smooth and timely implementation of the project	Adequate water supply and sanitation within the project area	Communication gap	District- and block-level meetings
State Environment Impact Assessment Authority	Issuing environmental clearance for development projects	Secure environmental clearance for the project	Compliance with statutory rule	Lack of coordination among key stakeholders in the implementation of development activities	Communication of the necessary information to the authority (as and when required)

continued on next page

Table A3 *continued*

Stakeholder	Roles and Responsibilities	Why the Stakeholder Should Be Involved	Motivating Factors[a]	Constraining Factors[b]	How the Stakeholder Should Be Involved[c]
SLSSC	Approving new water supply schemes	Comply with SLSSC statutory guidelines	Implementation of various schemes in the field	Limited availability of funds	State-level SLSSC meeting
SWSM	Responsible for the following: - providing policy guidance; - merging water supply and sanitation activities, including special projects; - coordinating with various state government departments and other partners in relevant activities; - monitoring and evaluating the physical and financial performance and management of water supply and sanitation projects; - integrating communication and capacity development programs for water supply and sanitation; and - maintaining the accounts for the program fund and the support fund, and carrying out the required audits	Frame water and sanitation policy and/or guidelines	Implementation of various schemes in the field	None	General water safety planning meetings
Water and Sanitation Support Organization	Responsible for the following: - providing policy and guidelines to facilitate the water safety planning process in the state; - committing funds for water safety planning activities; - monitoring and supervising water safety planning–related activities in the state; - providing training, IEC, laboratory, Water Quality Monitoring & Surveillance Program, etc., services;	Frame water and sanitation policy and/or guidelines	Implementation of different schemes in the field	None	SWSM general meetings

continued on next page

Table A3 *continued*

Stakeholder	Roles and Responsibilities	Why the Stakeholder Should Be Involved	Motivating Factors[a]	Constraining Factors[b]	How the Stakeholder Should Be Involved[c]
	- developing a state-specific IEC strategy for reform initiatives in water and sanitation; - developing the capacity of functionaries at all levels to address the need for sustainability in water and sanitation; - advocating conventional and traditional water conservation and rainwater harvesting; and - doing action research in various aspects of sanitation, including new technologies, impact of provision of sanitation facilities on health indicators, and IEC strategies				
West Bengal Disaster Management Department	Establishing the necessary systems, structures, programs, resources, capabilities, and guiding principles for reducing disaster risks, and preparing for and responding to disaster and threats of disaster in the state	Assess risks and ensure disaster preparedness	Minimized and/or managed risks associated with disaster	Communication gap	State- and district-level meetings
West Bengal Police	Ensuring safety on the Ganges River and protecting the river	Protect intake points from miscreants and/or intentional damage	Safety and security of the intake point Traffic management, maintenance of law and order	Gaps in coordination, less-than-timely communication	State-level meetings
West Bengal Pollution Control Board	Providing environmental monitoring and pollution control services	Protect raw water sources; ensure source quality and sustainability	Statutory mandate in place	Open disregard of the board's norms and guidelines by different stakeholders	State- and district-level meetings

continued on next page

Table A3 *continued*

Stakeholder	Roles and Responsibilities	Why the Stakeholder Should Be Involved	Motivating Factors[a]	Constraining Factors[b]	How the Stakeholder Should Be Involved[c]
Zilla parishad	Overseeing development work at the district level	Ensure smooth and timely implementation of the project	Adequate water supply and sanitation within the project area	Lack of or low awareness	District- and block-level meetings

ADB = Asian Development Bank; IEC = information, education, and communication; NOC = no-objection certificate; O&M = operation and maintenance; PWSS = piped water supply scheme; SLSSC = State Level Scheme Sanctioning Committee; SWSM = State Water & Sanitation Mission; VWSC= village water and sanitation committee; WBDWSIP = West Bengal Drinking Water Sector Improvement Project; WSP = water safety plan.

[a] Motivating Factors refer to factors that may motivate the stakeholder to adopt a safe system.

[b] Constraining Factors are factors that may demotivate the stakeholder and inhibit the adoption of a safe system.

[c] How the Stakeholder Should Be Involved refers to the interaction mechanism, e.g., consultations, meetings, sensitization workshops, etc.

Source: Prepared by the consulting team, 2019.

Members of Women's Self Help Group during an interaction with ADB team in Kalicharanpur Village, District Purba Medinipur.

Appendix 4: Checklist of Information about the Water Supply System to Be Compiled

Table A4: Some Information to Consider When Describing the Water Supply System[a]

Context and Institutional Framework	Identify contextual information about the system—political, economic, demographic, technological, environmental, social, and cultural factors that may influence water safety planning. Investigate any known health problems in the community or region and their possible relevance to water quality. A schematic summarizing the institutional framework of the sector and explaining how it works would be very helpful. Also collect enough data on quality standards, certification and audit requirements, and other legal and regulatory aspects related to the control of elements of the water supply management system, as well as laws that apply to public health and the environment.
General Information about the Water Supply System	A general overview of the system would provide useful information, including the water supplier (name, area of intervention, number of staff, governance model), population served, number of connections (by type), service level(s) (e.g., hours of operation, pressure), management systems, uses of water, volume of water demand and supply, water losses, and history of water quality issues. Summarize known persistent problems relating to water quality, continuity and quantity (reliability), accessibility, and management and operation, and be as specific as possible. Also state whether faults and deviations are followed up.
Climate	Identify changes in weather or other seasonal conditions, especially those linked to rainfall, drought, flood, salinity, and temperature variations.
Current and Required Quality of Delivered Water	Summarize the present quality of delivered water (at delivery points or at the customer's point of use). Highlight any differences between the required and the actual quality. Answer the following questions: ■ What are the legal operational requirements for water quality? ■ How has the quality of water for human consumption evolved over the years? Trace the trend over the last 5 years, if possible. ■ How many analyses of parametric values and percentage of compliance, per parameter, are done yearly? ■ What specific type of water quality control is exercised? How? When? Where? State the operational limits. ■ Is there a water quality control plan for supply and service activities (hygiene, repairs, etc.)? ■ What type of health surveillance is performed by the health authority? Have nonconformities been found? ■ Have users complained about water quality?
Catchment/Abstraction	Information about the following should be collected: capacity of the source in relation to demand; protection measures applied; developments in the catchment that may affect quantity and quality; summary of raw water quality (preferably over a long period and with statistical breakdowns); known water quality problems; and nature of contamination from sources of pollution like sewage treatment plants, urbanized areas, industries, and mining activities upstream of intake, and raw water delivery to treatment plants (if substantial distances are involved). Collect available aerial photographs, maps, planning schemes, land use maps, and policy documents, and consider answering the following questions: **Catchment** ■ What are the main hydrologic characteristics (with statistics) of the water source(s) (e.g., quantity and quality)? ■ What are the main catchment characteristics, including land use (e.g., household, sanitation, industry, agriculture, wildlife)?

continued on next page

Table A4 *continued*

Catchment/Abstraction *(continued)*	■ Is the catchment area delimited by protection perimeters? Have sources of contamination been identified? ■ Is the catchment area inspected periodically? ■ Have existing and future activities in the catchment area been identified? ■ Are there seasonal or weather variations? If yes, what is the impact of these variations on the quality and quantity of the water source(s)? **Abstraction** ■ What is the type of water source (surface water, groundwater)? How much water is abstracted? Where? How is abstraction performed? What is the flow rate? Level and depth? Hydrogeological impact? ■ What is the abstraction capacity and/or flow? ■ What is the abstraction infrastructure made of, and how old is it? ■ Are there alternative sources of water? ■ Is the abstraction licensed? ■ Are the water sources monitored (with respect to flow rates, levels, water quality, etc.)? ■ Have water quality problems been identified at the point of origin? ■ Have shortage problems been identified? ■ Is preventive maintenance performed? ■ Has the risk of cryptosporidium been identified? ■ What human activities take place near the abstraction point? ■ Are there protective measures around the abstraction area (e.g., fencing, grating)? ■ How sensitive is abstraction at the planned site to tidal (and salinity) intrusion?
	Prepare a simplified schematic of the system of major infrastructure downstream of the treatment plant, showing service reservoirs, transmission and trunk mains, booster stations, secondary service mains, zoning of supply, and any additional treatment (e.g., booster chlorination stations). Complement the schematic with a brief but informative description of the following: **Storage** ■ How many reservoirs are there? What is their capacity and type? What are they made of? And how old is the infrastructure? ■ How often are tanks inspected and rehabilitated? Is there a cleaning and disinfection plan for the reservoirs? ■ What kinds of problems have occurred in the reservoirs? ■ Does the entity have the means to repair the reservoirs when needed? ■ Are the storage tanks protected (e.g., rainproof with gutters)? ■ Are there screens on ventilation and overflows to prevent vermin and animal entry? ■ Is there adequate protection and/or security on storage tanks, with locked gates and hatches? ■ Are there separate inlets and outlets at varying heights on opposite sides of tanks to promote good mixing? **Distribution** ■ What is the size of the distribution network (in kilometers)? ■ Is the distribution constant or intermittent? Are there areas with different service levels (e.g., fully plumbed, yard taps, public tap stands)? ■ State the network characteristics (gravitational or pressurized, pressure stages, reducing valves, network ends, number of lifting stations, number of overpressures, etc.). ■ Is the distribution network divided into measurement and control zones? How does water flow through the system within each zone? In cross-connections between supply zones and between areas of different population density? ■ Is there information about pressure (constant or intermittent) and flow rates in the distribution network? ■ Is there an updated network registration? In what format? Is there a geographic information system map of the network?

continued on next page

Table A4 *continued*

Storage and Distrib ution *(continued)*	▪ Is there information about the type, material, average age, and state of repair of the distribution network and its components and/or accessories (extensions, valves, suction cups, hydrants, etc.)? ▪ Are actual network losses counted? Have the critical points in the network (in terms of losses, breakages, etc.) been identified? ▪ How often is the network inspected (e.g., to detect illegal connections)? What type of problems occur most often in the distribution network? ▪ Describe the physical condition of the components of the distribution system (pipe age, pipe diameter, pipe length and jointing, pipe materials, age of the service reservoirs or supply tanks, age of major valves). ▪ Give the location and description of known problem spots susceptible to contamination within the network and distribution system. Are there known leakage issues near sewerage mains or other sources of contamination? ▪ Is there a backflow prevention mechanism? ▪ Is there a cleaning and disinfection plan for the distribution network? ▪ How often is maintenance, renovation, rehabilitation, or replacement work done on the network? Is there an investment plan for renewing the distribution network? Does the entity have the ability to make the appropriate repair and/or renewal interventions in the network? ▪ Is there secondary disinfection, and if so, are chlorine residuals in critical points in the system monitored and recorded?
Treatment	Provide information about the treatment configuration, treatment process (a brief description), age of plant, known design faults, chemicals added, and treated water quality (a summary, preferably covering a long period and with statistical breakdowns). Consider answering the following questions: ▪ Identify existing treatment processes and treatment capacity. Analyze treatment efficiency and/or effectiveness. ▪ What chemicals and materials are used for treatment? What is their availability? What is the quality of available chemicals? How are the chemicals stored? ▪ Is the water disinfected? If so, what methods and disinfectants are used? Is there sufficient disinfectant (e.g., chlorine) contact time for proper disinfection? ▪ Indicate the year of construction of the facility and the condition of the treatment infrastructure, and state whether rehabilitation, renewal, or replacement of treatment infrastructure and/or processes has already taken place. ▪ Do purchased chemicals undergo quality control checks? Do they meet the respective quality standards? ▪ Identify the type of monitoring that is performed during the treatment process (What? Where? How? When?). What are the operational and critical limits? What actions are taken when those limits are exceeded? ▪ Identify the types of problems that have already occurred. Compile infor mation about faults and occurrences. ▪ Indicate the type of maintenance performed on the infrastructure. ▪ Is water quality monitored? How? When? Where? ▪ Are the treatment plant operators trained? Are there minimum competency standards, and do the operators meet those standards?

continued on next page

Table A4 *continued*

Delivery Points, Intended Uses and Users of the Water, and Customer Practices	Consider the following issues (if applicable): ■ Water uses: - What are the current uses of water (e.g., drinking, food preparation, personal hygiene, clothes washing, domestic livestock, vegetable farming, fish market) and the future needs? Specify quantity and quality. - What are the types of users, including commercial users (e.g., homes, hotels, guesthouses, institutions, workshops, small industry), and how many are there of each type? - Are there vulnerable groups or groups with special needs within the population, including the infirm or sick and the aged? - Are there hospitals and schools? ■ What material is used for domestic pipe work, and how old is it? ■ Household collection and storage practices: - Do households treat and store water? - By what means? - How is water collected and transported? ■ Operation and monitoring: - Are standposts and house connections inspected? - Is water quality tested? How? How often? By whom? ■ What consumer education is in place for water use and how is this conveyed? How are consumers notified of potential contamination? Are consumers aware of regulatory requirements for drinking water quality (e.g., drinking water standards)? ■ How is wastewater handled?

[a] Adapted from World Health Organization South-East Asian Regional Office. 2016. *Capacity Training on Urban Water Safety Planning: Training Modules*. Delhi.

Source: Prepared by the consulting team, 2019.

Appendix 5: Environmental Legislation, Strategy, and Guidance Documents

Table A5: Environmental Legislation and Guidance Documents

Application	Document
Nationwide	**National Environment Policy (NEP), 2006** The NEP is a comprehensive guiding document in India for all environmental conservation programs and legislation of the central, state, and local governments. The dominant theme of this policy is promoting the betterment of livelihoods without compromising or degrading environmental resources. The policy also advocates collaboration among stakeholders to harness potential resources and strengthen environmental management.
Nationwide	**Water (Prevention and Control of Pollution) Act, 1974 (amended 1988), and its Rules, 1975** This ordinance was passed by central and state boards to prevent and control water pollution, and to maintain or restore the wholesome properties of water.
Nationwide	**Environment (Protection) Act, 1986** Covers the protection and improvement of the environment and the prevention of hazards to human beings, other living creatures, plants, and property.
Nationwide	**Coastal Regulation Zone (CRZ) Notification, 2011** This supersedes the CRZ Notification issued in 1991 and is intended to ensure livelihood security to fisher communities and other local communities living in the coastal areas, to conserve and protect coastal stretches and their unique environment, and to promote sustainable development considering natural hazards and sea-level rise due to global warming. Coastal stretches are declared CRZs, and new construction and industrial activities are restricted.
Nationwide	**Environmental Impact Assessment (EIA) Notification, 2006** Issued under the Environment Protection Act, 1986, the EIA Notification of 2006 (replacing the EIA Notification of 1994) sets out the requirement for environmental impact assessment.
Nationwide	**Environment (Protection) Rules, 1986 (including amendments)** These rules specify the following: - standards for emissions or discharge of environmental pollutants, - prohibitions and restrictions on the location of industries, - procedure for taking samples and submitting samples for analysis, - prohibitions and restrictions on the handling of hazardous substances in different areas, and - procedure for submitting environmental reports
Nationwide	**Wetland (Conservation and Management) Rules, 2010** The rules provide a regulatory mechanism for the protection of wetlands and the restriction of certain activities within wetlands. They apply to protected wetlands notified under the rules (including Ramsar sites, wetlands in eco-sensitive zones and/or United Nations Educational, Scientific and Cultural Organization (UNESCO) sites, wetlands in high altitudes, etc.). Activities such as the following are regulated: water withdrawal and/or diversion, treated effluent discharge, dredging, repair of existing infrastructure, and building and construction.
Nationwide	**Major Port Trusts Act, 1963–Kolkata Port Trust** Prior permission of Kolkata Port Trust Board is required for any construction, mooring, reclamation, etc., in port limits and port approaches.

continued on next page

Table A5 *continued*

Application	Document
State level	**West Bengal Ground Water Resources (Management, Control and Regulation) Act, 2005** The West Bengal State Level Ground Water Resource Development Authority was established under this act to manage, control, and regulate indiscriminate extraction or use. The State Water Investigation Directorate is its functional organ for implementing the act. Permission from the authority is mandatory for the construction of groundwater extraction structures (operated by engine or motor-driven pump).
State level	**East Kolkata Wetlands (Conservation and Management) Act, 2006** This act is for the conservation and management of the East Kolkata Wetlands (EKW), spreading over 12,500 hectares in Kolkata and in the North and South 24 Parganas districts of West Bengal. The EKW Management Authority was created under this act to conserve the wetlands, make rules, enforce land use controls, and regulate all activities. Prior permission from the authority is required for any project activities in the notified area.
State level	**West Bengal Action Plan on Climate Change, 2012** Water resources are among the major components dealt with in detail in the action plan. The plan highlights regional variations in water availability, demand, quality, etc., considering likely changes in rainfall, temperature, blue water flow, green water flow, and green water storage. It suggests various regionwide strategies and an action plan for water resource management. The Public Health Engineering Department is one of the agencies responsible for implementing the action plan.
State level	**Guidelines for National Rural Drinking Water Programme, 2013**
State level	**Guidelines for Swachh Bharat Mission (Gramin), 2017**
State level	**Revised Guidelines of National Water Quality Sub-Mission, 2017**
State level	**Guidelines for Solid and Liquid Waste Management in Rural Areas, 2017**
State level	**Guidelines for Open Defecation Free Sustainability, 2016**
State level	**Manual for the Preparation of Detailed Project Report for Rural Piped Water Supply Schemes, 2013**
State level	**Toolkit for the Preparation of a Drinking Water Security Plan, 2015** Guidelines drawn up with the support of the World Bank and the Ministry of Drinking Water and Sanitation.
State level	**Operation and Maintenance Manual for Rural Water Supply, 2013**
State level	**Uniform Drinking Water Quality Monitoring Protocol, 2013**
State level	**Technological Options for Solid and Liquid Waste Management in Rural Areas, 2015**
State level	**Implementation Manual on National Rural Water Quality Monitoring and Surveillance Programme, 2004**
State level	**Sanitation and Hygiene Advocacy and Communication Strategy Framework 2012–2017, 2012**
State level	**Drinking Water Advocacy and Communication Strategy Framework 2013–2022, 2013**
State level	**Manual on Water Supply and Treatment, 1999**
State level	**Manual on Sewerage and Sewage Treatment Systems (Revised), 2013**
State level	**Manual on Operation and Maintenance of Water Supply Systems, 2005**

continued on next page

Table A5 *continued*

Application	Document
State level	**Manual on Rain Water Harvesting and Conservation, 2002** Prepared by the Central Public Works Department.
State level	**IS 10500: Drinking Water Specification (Second Revision), 2012**
State level	**IS 1172: Code of Basic Requirements for Water Supply, Drainage and Sanitation (Fourth Revision), 2002**
State level	**Air (Prevention and Control of Pollution) Act, 1981 (amended in 1987)**
State level	**Solid Waste Management Rules, 2016**
State level	**Wildlife (Protection) Act, 1972**
State level	**Environmental Impact Assessment Notification, 2006**

Source: Prepared by the consulting team, 2019.

Heavy reliance on groundwater puts most of the rural population in West Bengal at risk from arsenic and fluoride contamination that can lead to health problems including cancer and bone diseases.

Appendix 6: Brief Description of the Bulk Water Supply Systems for Bankura and North 24 Parganas Districts

Bankura District

Bankura District is located in the western part of the state of West Bengal. It is bounded by the districts of Bardhaman on the north, Purulia on the west, and Paschim Medinipur on the south.

Bankura has hilly streams originating in the highlands in the east and flowing from the northeast to the southwest. The Damodar River, forming the northern border with Bardhaman, flows into the district. The Dwarakeshwar, the Shilabati, and the Kangsabati are other major rivers, while the Sali and the Gandheshwari are important tributaries of the Damodar and Dwarakeshwar rivers, respectively. A dam across the Kangsabati in Khatra Community Development Block was built under the Kangsabati Reservoir Project in 1956.

The area has a tropical dry climate. The average annual temperature varies from 10°C to 46°C, while the annual mean temperature is around 25°C.

Summers are hot and humid and, during dry spells in May and June, the maximum temperature often exceeds 40°C. Winter tends to last for only 2.5 months, with seasonal lows dipping to 5°C–6°C between December and January.

Rains brought by the Bay of Bengal occur between June and September. The average annual rainfall in the district has varied from a low of 917 millimeters (mm) in 2010 to a high of 1,800 mm in 2014.

The total area of Bankura is 106,882 square kilometers (km^2). The population of 113,596,674 (according to the latest census in 2011) makes the district West Bengal's third least populated (after Alipurduar and Purulia), with a population density of 523 persons/km^2.

Bankura has 22 *panchayat samitis* (block-level governance units), with 190 *gram panchayats* (village-level governance units), consisting of 3,823 villages and 6,638 habitations.

The total number of urban centers is 12, where 3 are municipalities (Bankura, Bishnupur, and Sonamukhi) and the remaining 9 are census towns (Barjora, Beliatore, Ghutgarya, Jhanti Pahari, Khatra, Kotulpur, Ledisol, Raipur Bazar, and Simlapal).

The economy of Bankura district is predominantly agrarian. The crop pattern is tilted heavily toward paddy cultivation using traditional practices. The unconducive topography, very small landholdings, poor irrigation coverage, and low water retention capacity of soil, among other factors, offer limited scope for farm mechanization. Around 70% of the people in Bankura still depend on agriculture for their daily livelihood. Of the net cultivated area of the district, 42% are in drought-prone or near-drought-prone regions, with very serious impact on per capita consumption in the locality.

The handloom industry and pottery engage the largest number of people in the nonfarm sector and, hence, are important in Bankura district. The district is well known for the Baluchari *sari* (traditional garment) and the artistic excellence of its pottery products.

In Bankura district, 41.52% of families live below the poverty line (2011 Census of India). This proportion is much higher than the average state figure. Poverty is unevenly distributed in the district, with economically backward areas located mostly in its western and southern areas. The malnutrition problem in Bankura district is multidimensional. On average, every other child in the district is moderately or severely malnourished.

On public health issues, diarrhea occurs throughout the year in almost all the blocks in Bankura district. Sporadic outbreaks occur mainly in late summer because of water scarcity. Cases in the district have steadily risen. However, the apparent rise is due to the fact that vigilance has increased and reporting has improved. Malaria is also a major public health problem in the district. The causes of the increase in malaria cases include resistance to DDT (dichloro-diphenyl-trichloroethane) and the treatment drug chloroquine, inaccessibility of affected areas, lack of personal protection, and lack of a behavior change communication campaign to control the vector-borne disease. Another vector-borne disease that is endemic in Bankura district is filariasis. All the blocks are more or less affected, and the district lacks a rural filaria control program. With the increasing nonavailability of water and the lack of suitable drainage, vector-borne diseases are bound to multiply.

Ongoing Water Supply Schemes

Seven water supply schemes are ongoing in Bankura. Of these, five are based on subsurface sources, and one is dependent on groundwater.

The Public Health Engineering Department (PHED) of West Bengal appointed consultants, Development Consultants Private Limited, to study and report on the status of existing schemes in Bankura district. The consultants mainly analyzed the piped water supply schemes implemented in the last few decades as well as the ongoing ones. In all, seven selected schemes in the three selected blocks of Gangajalghati, Mejia, and Indpur were analyzed and reported on by the consulting team. The findings of the team are summarized in Table A6.1.

Table A6.1: Status of Piped Water Supply in Selected Blocks

Name of Scheme and Block	Year Commissioned	Source of Water	Description of Water Supply System	Comments and/or Suggestions
Indpur Zone-3 Water Supply Scheme	More than 25 years old	Tube well in the bed of Dwarakeswar River	TW–CWR–CW pumps–distribution through PSPs	Not advised to go for renovation and modernization of the scheme.
Goaldanga Water Supply Scheme—Indpur Block	2016	Tube well installed in the bed of Shilabati River	TW–CWR–disinfection–CW pumps–OHT–distribution	Water is supplied to eight *mouzas* (administrative districts)—six fully and two partially—through stand posts. Distribution network to be modified for household connection.
Gangajalghati (Zone 1 and 2) Water Supply Scheme	2005	Ganga Dam on Sali River	Intake–plate settler–chlorination–CWR–CW pump–pressure filter–OHT–distribution	Distribution network to be modified for household connection. Scheme to continue.

continued on next page

Appendix Table A6.1 *continued*

Name of Scheme and Block	Year Commissioned	Source of Water	Description of Water Supply System	Comments and/or Suggestions
Charradih Water Supply Scheme—Gangajalghati Block	1969	Tube wells installed in the bed of Shali River	Infiltration gallery–CWR– CW pumps–OHT–distribution	Not advised to go for renovation and modernization of the scheme. Too old a project for revival.
Mejia Water Supply Scheme–Phase I	2001	Three tube wells with submersible pumps installed in the bed of Damodar River	TW with submersible pumps–CWR–CW pumps–distribution through PSPs	Physical condition of CW pumps and associated pipework is poor. Distribution system to be redesigned to meet household connection needs. Scheme to continue.
Mejia Water Supply Scheme–Phase II	Not yet commissioned	Three tube wells with submersible pumps installed in the bed of Damodar River	TW with submersible pumps–CWR–CW pumps–distribution through PSPs	Distribution system to be redesigned to meet household connection needs. Scheme to be commissioned with modifications.
Bhara Water Supply Scheme—Mejia Block	2015	Tube wells with submersible pumps installed in the bed of Damodar River	TW with submersible pumps–CWR–CW pumps–distribution through PSPs	Distribution system to be redesigned to meet household connection needs. Scheme to continue.

CW = clear water, CWR = clear water reservoir, OHT = overhead tank, TW = tube well, PSP = power steering pump.
Source: Prepared by the consulting team, 2019.

Drinking Water Sources and Quality issues

The majority of the rural areas in Bankura district have been provided with water supply systems, mostly through provision of hand pumps. However, because of the low-retaining aquifers and hard-rock areas, many of these hand pumps either go dry or have low yield during the hot summer season. Hence, a sizable part of the population of the district goes without safe drinking water and is compelled to recourse to unsafe sources. In Gangajalghati, Mejia, Indpur, and part of Taldangra Block, machine boring is the only way of installing tube wells because of the hard-rock areas. However, the quality of most of these sources is uncertain: yield is very low and comes mostly from poor water-bearing formations.

Besides water scarcity, the doubtful quality of water, particularly in view of the high concentrations of fluoride and iron in groundwater, poses a serious problem for this district, according to the Central Ground Water Board, Ministry of Water Resources. Groundwater in Taldangra, Indpur, and Gangajalghati blocks (among many other blocks in the district) is sporadically affected by high concentrations of fluoride (>1.5 milligrams per liter [mg/l]).

Iron in groundwater has also been found to be present in concentrations that are quite high (up to 9.5 mg/l). Though the iron content of drinking water may not affect the human system, prolonged accumulation of iron in the body may result in hemochromatosis, a disease that causes tissue damage.

Thus, a water cluster source, as a more promising and sustainable source, should be explored while available options in the proximity of the affected blocks were taken into consideration in framing the new proposals.

New Piped Water Supply Schemes

The piped water supply systems under the West Bengal Drinking Water Sector Improvement Project (WBDWSIP) will provide a minimum of 70 liters per capita per day (lpcd) of potable water through metered household connections on a 24/7 basis to each household in the selected blocks and villages in Bankura district, and potable bulk water at the prescribed national standards to en-route habitations. The distribution systems, designed on a district metering area basis, will reach the household level, including community and government institutions such as schools and *anganwadis* (community-based child care and development center), and will be equipped with district meters and domestic water meters. Both bulk and distribution systems will be integrated with state-of-the-art smart water monitoring (SWM) and SWM tools, including supervisory control and data acquisition (SCADA) and geographic information systems. Where feasible, bulk water supply systems will be interconnected by means of a grid.

The PHED will be the executing agency, responsible for planning, designing, and carrying out the project, and monitoring the implementation of the bulk water systems up to the boundary of the *gram panchayats*. It will conduct operation and maintenance (O&M) activities at the intake source, the treatment plant, the pumping mains, and the overhead reservoir (OHR) sites.

Panchayats and other local bodies will be involved in O&M activities for distribution zones in their respective jurisdictions, as the provision of house connections is part of the scheme. The annual O&M cost of the scheme will mostly be covered by a suitable tariff and a one-time deposit of connection charges to be decided by the competent authority.

The Durgapur Barrage, along with its upstream area, and the Mukutmanipur Dam on the Kangsabati River are the only potential surface and subsurface abstraction points that were considered for the two water supply schemes for the two delivery clusters: the Indpur and Taldangra blocks, and the Mejia and Gangajalghati blocks.

Total water demand for the Indpur and Taldangra blocks for the *mouzas* (administrative districts) under the project has been estimated at 44 million liters per day (MLD) in 2050. Raw water is expected to be made available from Mukutmanipur Dam through the construction of an intake chamber and a water treatment plant (WTP). The dam, located around 19 kilometer (km) from Indpur, is expected to meet the total demand of Indpur Block (22.1 MLD) and Taldangra Block (21.8 MLD).

The intake for this scheme will be in Satsol Mouza in Khatra Block, and the treatment plant will be in Loadihi Mouza in the same block. The plant will be a conventional WTP. Treated water from the clear reservoir will be pumped to the intermediate ground-level service reservoir (GLSR-I) proposed for construction near Gobindapur Village.

Treated water stored in this reservoir will then be pumped to a ground-level storage reservoir–cum–pumping station in Indpur Block (GLSR-II) and in Taldangra (GLSR-III). At the same time, the treated water will also be fed to adjoining overhead tanks along the way. Treated water from storage reservoirs GLSR-II and GLSR-III will be further pumped to overhead tanks at the tail end of the blocks. These overhead tanks will be built at appropriate locations to serve all the habitations within the block by means of a gravity-based supply system.

The entire water supply distribution network will include comprehensive household service connections and water metering.

Table A6.2: Indpur–Taldangra Piped Water Supply Scheme in Bankura District under the West Bengal Drinking Water Sector Improvement Project

Item	Details
Location	Indpur and Taldangra blocks
Number of *mouzas* (administrative district) (2011 census)	Indpur : 202 Taldangra : 145
Census population (2011), including weaker section	Indpur : 156,522 Taldangra : 147,893
Design population (2050)	Indpur : 201,247 Taldangra : 188,040
Command area	Indpur : 300.2 km^2 Taldangra : 349.7 km^2
Number of schemes	1
Number of zones	Indpur : 20 Taldangra : 24
Per capita service level	Rural and urban: 70 lpcd (including 10% through standposts for public distribution)
Daily water demand	Raw water: 44.0 MLD Indpur : 22.1 MLD Taldangra : 21.8 MLD
Source of water	Mukutmanipur Dam
Quality aspect of source	Water will require conventional treatment before distribution.
Quantity aspect of source	Adequate to meet future demand
Location of intake	Satsol Mouza in Khatra Block
Tentative location of treatment plant	Loadihi Mouza in Khatra Block
Daily hours of supply to consumers	24 hours
Pumping period	20 hours
Raw water main	DI (K9)/MS pipe of suitable size with necessary wrapping and coating

DI = ductile iron, K9 = a specific thickness class of pipe, km^2 = square kilometer, lpcd = liter per capita per day, MLD = million liters per day, MS = mild steel.

Source: Prepared by the consulting team, 2019.

The Gangajalghati Community Development Block has an area of 366.47 km^2. According to the 2011 census, the Gangajalghati Block had a total population of 180,974, all of which was classified as rural. The Mejia Community Development Block had a total population of 86,188 in 2011, all classified as rural. Mejia and Gangajalghati blocks will get their water from Durgapur Barrage on the Damodar River. Total raw water intake will be 34 MLD. Of this, 12 MLD will be for Mejia and 22 MLD for Gangajalghati Block. The intake will be in Nutangram Mouza in Barjora Block, and the treatment plant will be in Basudebpur Mouza, Gangajalghati Block. Seventy-four *mouzas* in Mejia Block with a population of 86,188 in 2011, and 165 *mouzas* in the Gangajalghat block with a 2011 population of 180,974, will be provided with 70 lpcd.

Table A6.3: Mejia–Gangajalghati Piped Water Supply Scheme in Bankura District under the West Bengal Drinking Water Sector Improvement Project

Item	Details
Location	Mejia and Gangajalghati blocks
Number of *mouzas* (administrative districts) (2011 census)	Mejia : 74 Gangajalghati : 165
Census population (2011), including weaker section	Mejia : 86,188 Gangajalghati : 180,974
Design population (2050)	Mejia : 139,911 Gangajalghati : 278,347
Command area	Mejia : 162.87 km^2 Gangajalghati : 366.47 km^2
Number of schemes and zones	1
Number of zones	Mejia : 9 Gangajalghati : 25
Per capita service level	Rural and urban: 70 lpcd (including 10% through standposts for public distribution)
Daily water demand	Raw water : 34 MLD Mejia : 12 MLD Gangajalghati : 22 MLD
Source of water	Surface water of Durgapur Barrage
Quality aspect of source	Water will require conventional treatment before distribution.
Quantity aspect of source	Adequate to meet future demand
Location of intake	Nutangram Mouza in Barjora Block
Tentative location of treatment plant	Basudebpur Mouza in Gangajalghati Block
Daily hours of supply to consumers	24 hours
Pumping period	20 hours
Raw water main	DI (K9)/MS pipe of suitable size with necessary wrapping and coating
Clear water rising main	DI (K9)/MS pipe of suitable size with necessary wrapping and coating

DI = ductile iron, K9 = a specific thickness class of pipe, km^2 = square kilometer, lpcd = liter per capita per day, MLD = million liters per day, MS = mild steel.

Source: Prepared by the consulting team, 2019.

North 24 Parganas District

North 24 Parganas District lies between Nadia District on the north and the Bay of Bengal on the south. Much of its eastern boundary is shared with Bangladesh. On the west, it is bounded by Kolkata and the Hooghly River. Its southwest neighbor is the district of South 24 Parganas.

North 24 Parganas is West Bengal's most populous district (and India's second most populous), with a population of 10,009,781 (2011 census) spread over an area of 4,094 km^2, for a population density of 2,463/km^2. Rural and urban community development blocks in the district have a total population of 5.01 million, organized into 1.17 million households.

The urban part of the district comprises 107 urban centers—27 municipalities, 78 census towns, 1 cantonment board, and 1 industrial township. There are 22 *panchayat samitis*, 200 *gram panchayats*, and 1,527 villages (9 of these uninhabited) in the 22 community development blocks (2011 census).

The district is divided into three zones: (i) a highly industrialized north, (ii) a moderately industrialized northeast, and (iii) a highly agricultural northeast. Sixty-six percent of the total area of the district is cultivable land. Agriculture is still a major source of livelihood in rural North 24 Parganas; but, with urbanization, the land use pattern in the district is rapidly changing and the area under cultivation is shrinking. In many parts of the district, orchards and brick kilns are replacing croplands. Also, while much of the land in the Sunderban area is still used for pisciculture, particularly for shrimp cultivation, agricultural land upstream of the Ichamati has experienced extensive flooding following the loss of natural flow of water through this river. Over-silting, coupled with encroachment on the Ichamati, has worsened the problem.

The main rivers in the district besides the Ichamati are the Benti, Bidyadhari, Borokalagachi, Dansa, Gaourchrar, Haribhanga, Hooghly (also known as the Ganga), Kalindi, and Raimangal rivers.

The climate can be categorized as subtropical, with a monsoon regime. In general, it is hot and humid between March and October. Cooler dry weather sets in by mid-November and lasts until about mid-February. Average temperatures range between a low of 13°C in January and a high of 36°C in April or May. Limited rainfall occurs from the latter part of February to the end of May. The main rainy season coincides with the southwest monsoon from the first week of June to the end of September.

Water Quality Status

Groundwater is mildly alkaline, with pH values ranging between 7.5 and 8.2. Total hardness, expressed in terms of mineral (calcium carbonate) content, ranges from 140 to 670 mg/l. Iron content is generally above the national permissible limit of 0.3 mg/l in all the blocks, ranging from 1.23 to 18.10 mg/l; but, in a few places, it is lower: 0.09–0.56 mg/l.

Groundwater in the area is typically of the bicarbonate type. The chloride content is low (18–234 mg/l) in the northern and central parts of the district. In the southern and southeastern parts (Basirhat, Haroa, Hasnabad, Hingalganj, Minakhan, and Sandeshkhali), the upper aquifers are brackish to saline, but freshwater aquifers underlie the saline aquifers.

Shallow aquifers within a depth range of 20–80 meters below ground level (mbgl) show arsenic concentrations above the permissible limit (0.01 mg/l) in drinking water occurring in all the 22 blocks in the district. Deeper aquifers down to a depth of 350 mbgl are arsenic-free.

On surface water quality, the Hooghly River is now the major water source for a number of water treatment plants in West Bengal. Typical water quality results indicate high coliform contamination, which is probably due to discharge from habitations on both sides of the river. Turbidity reaches 500 nephelometric turbidity units (NTU) during the monsoon. There have been no reports of chemical contamination, though this does not mean that none exists.

Present Coverage of Water Supply

North 24 Parganas has a predominantly urban profile, with almost 50% of the people living in the 27 municipalities. According to the 74th Constitutional Amendment Act, urban local bodies are supposed to arrange their own water supply systems, while the state PHED is responsible for rural water supply in the blocks and the census towns within the blocks. Depending on the mode of delivery, water supply in the rural areas of North 24 Parganas can be categorized into two groups: (i) piped water supply schemes, and (ii) spot source–based schemes (like dug wells, shallow wells, and deep bore wells) fitted with hand pumps. Of the rural habitations in the district, 4.9% are served by surface water systems and alternative sources; a substantial proportion still get their supply from groundwater-based spot sources.

According to data from PHED sources, there are 181 commissioned piped water supply schemes (PWSSs) in this district, and work on 32 other schemes is in progress. Of the 181 commissioned schemes, 170 (94%) are based on groundwater and 11 on surface water. Around 77 schemes (43%) are older than 12 years, and only 5 groundwater-based PWSSs were approved in the last 7 years.

These data suggest the slow and planned phaseout of the installation of groundwater-based schemes, as recommended in the arsenic master plan drawn up by the state government, and a preference for surface water–based schemes.

Groundwater has been the only source for most of the piped water supply schemes in the rural areas. Sixty-four percent of these schemes are small, each one covering 10 or fewer habitations.

The typical groundwater-based scheme comprises a bore well and pumps, supplying water to households after disinfection through chlorination.

Where surface water sources are too far away, groundwater-based schemes equipped with an arsenic and iron removal plant are an option in the PHED's arsenic master plan.

For some arsenic-affected areas in the North 24 Parganas District, the PHED has initiated the use of natural or constructed ponds close to habitations for storing rainwater. Water from the ponds is withdrawn by means of a double-stroke hand pump or electrical pump and treated in a horizontal-flow roughing filter with several compartments, followed by slow sand filtration.

At present, the only major surface water–based piped water supply scheme in operation in the district is the North 24 Parganas scheme. This scheme mainly covers the four blocks of Amdanga (partly), Deganga, Barasat–I, and Barrackpore–II, as well as a small portion of Basirhat–I. It was commissioned in two phases (2006, 2008) for a design population of 0.768 million (by 2025).

The raw water intake jetty is on the Hooghly River, near the mechanized brick factory in Palta, Barrackpore Municipality. A 34 MLD conventional treatment plant, with clariflocculation and rapid gravity filtration, supplies treated water to the command area through a 220 km-long transmission line via two booster stations to 14 overhead storage tanks. These cover 227 *mouzas* spread over an area of 369 km^2. The scheme, designed to supply 40 lpcd at the consumers' end, will have to be augmented to meet the PHED's Vision 2020 target of 70 lpcd. As the scheme is reaching half of its design life, some mechanical and electrical equipment may also have to be replaced.

Ongoing Piped Water Supply Schemes

At present, there are 32 PWSSs under implementation in the district, which comprise the following:

- 2 surface water–based PWSSs,
- 1 pond water–based scheme,
- 1 groundwater-based dual solar pump scheme,
- 1 reverse osmosis plant-based groundwater scheme, and
- 27 borewell groundwater-based schemes.

The establishment of surface water–based bulk water supply systems is proposed under the WBDWSIP to meet the water demand of the arsenic (groundwater)–affected blocks of Rajarhat and Haroa in North 24 Parganas and Bhangar–II in South 24 Parganas. These blocks are on the eastern side of the state capital, Kolkata.

The contract for this bulk water supply project (WBDWSIP/DWW/N24P/NCP/01) covers a water treatment plant, reservoirs, transmission mains, and pumping stations at existing GLSRs–cum–booster pumping stations in Haroa and Bhangar–II, including SCADA. Salient features of the project are presented below.

Water Source

Raw surface water is currently abstracted from the Hooghly River in Rani Debendra Bala Ghat near Bagbazar, and delivered through a transmission main (11.5 km long and 1,829 mm in diameter) to the WTP in Rajarhat.

The intake is functional and has the capacity to draw 100 million gallons of water per day (MGD) (or equivalent to 454 MLD). A transmission main has also been laid to bring raw water from the river to the New Town Water Treatment Plant in Kolkata.

The raw water intake at the WTP and the raw water transmission main to the WTP for the Haroa–Rajarhat–Bhangar II block scheme will be integrated with the existing water supply schemes but will be managed separately for the blocks covered under the project.

The quality of Hooghly River water at the intake point is fit for use in public water supply after appropriate treatment. The PHED did an extensive study to assess the quality of the river in different seasons. The pH of the raw water in the river ranged between 6.5 and 8.5. The biochemical oxygen demand of river water near the intake point was found to be around 3 mg/l or less. Dissolved oxygen ranged between 5.0 and 6.5 mg/l. The total coliform count was more than 5,000/100 milliliters (most probable number count).

The raw water is contained in five presettlement ponds adjacent to the WTP site in Rajarhat. Storage in the ponds facilitates natural sedimentation, thereby considerably reducing the suspended solids as well as turbidity. Turbidity in pond water after sedimentation is expected to vary between 40 and 100 NTU when raw water turbidity ranges between 100 and 900 NTU.

Water Treatment Plant

A new 22 MGD (or 100 MLD) water treatment plant will be built in New Town to cater to the command area of the North 24 Parganas subproject under the WBDWSIP. The water treatment plant, in principle, will comprise a rapid mixing unit, a clariflocculator or flocculation unit and plate settler, a rapid sand filter, and a disinfection facility. Chemical coagulation and flocculation (using alum or poly-aluminum chloride) will be applied. Liquid chlorine will be used for disinfection. A clear water reservoir (CWR) will be built to store the treated water.

A contact time of 30 minutes will be provided after disinfection. Booster chlorination will maintain residual chlorine within 2 mg/l at the consumers' end. WTP plant operators will be trained to operate and maintain the WTP properly. Experienced operators will be chosen to run the WTP.

A water quality monitoring and surveillance program will be implemented at the WTP. Water quality will be regularly monitored in the WTP laboratory, and performance will be evaluated at established intervals. Operational guidelines will be prepared for the WTP to facilitate efficient plant operation.

Bulk Water Supply System

Treated water from the proposed CWR and pumping station of the WTP will be pumped to a proposed CWR-cum-booster pumping station, with a capacity of 4,600 kiloliters (kl). From this point, treated water will be pumped to ground-level reservoirs (GLRs)-cum-booster pumping stations in the following blocks:

- Haroa (capacity: 3,200 kl),
- Bhangar-II (capacity: 5,000 kl), and
- Rajarhat (capacity: 1,000 kl).

Clear water transmission mains from the GLRs will convey the treated water to OHRs in the respective blocks for further distribution.

Classroom at Brij Mohan Tewary Girls High School in Nandigram, Purba Medinipur, West Bengal.

Appendix 7: Hazardous Events and Hazard Risk Assessment

Table A7: Sample Hazard Risk Assessment

Implementation Phase	Process Step	Component	Hazardous Event (X happens because of Y)		Hazard (microbial, M; physical, P; chemical, C; disruption of water supply)	Additional Information
			X = What Can Happen to the Water Supply	Y = How It Can Happen (i.e., cause)		
Design	Water source	Intake	Nonavailability of raw water	Improper selection of water source due to flaw in initial bathymetric survey followed by inaccurate location of intake point or unattended long-term morphological changes	Water supply disruption	Intake point should be selected judiciously.
Design	Water source	Intake	Nonavailability of raw water	Long-term changes in water level statistics at the river intake, causing flooding or drought, or elevated salinity	Water supply disruption at certain times during the year	Selection of intake point and design of water intake should be preceded by proper hydrological studies or coastal studies, including climate change assessment.
Design	Treatment	Water treatment plant	Inadequate water treatment	Improper design of treatment process of water due to the following: - lack of detailed assessment of raw water quality - inattention to seasonal variations in raw water quality	C	Water treatment plant design should be based on detailed and prolonged analysis of raw water quality data to ensure complete treatment of water.

Existing Control Measures (measures already in place to address hazard)	Are These Controls Effective?	Validation Notes (basis of control measure effectiveness decision)	Risk Assessment					Is Additional Control Needed?	If Yes, Proposed Controls (to be developed further in the improvement plan)
			Likelihood	Consequence	Risk Score (PxS)	Risk Level	Validation Notes		
Survey kept within scope of contractor's responsibilities	Partly	Work done under strict surveillance; report compared with previously available data	1	2	2	Low	PHED and DSISC should ensure survey accuracy.	Yes	Survey accuracy ensured by PHED and DSISC
Design studies within scope of contractor's responsibilities	Partly	Quality assurance review of work by relevant experts	3	5	15	High	The closer the intake is to the coastal area, the greater is the likelihood of such supply disruptions.	Yes	A forecast and early warning system established near coastal areas
Conventional type of water treatment plant proposed	Partly	Heavy metals, pesticides, etc., found in water may not be removed through conventional treatment.	2	3	6	Medium	If pesticides are present in the raw water, additional treatment may be required.	Yes	Detailed testing of raw water quality to determine the presence of heavy metals, pesticides, etc., and implementation of proper treatment measures

continued on next page

Table A7 *continued*

Implementation Phase	Process Step	Component	Hazardous Event (X happens because of Y)		Hazard (microbial, M; physical, P; chemical, C; disruption of water supply)	Additional Information
			X = What Can Happen to the Water Supply	Y = How It Can Happen (i.e., cause)		
Design	Transport	Transmission main, distribution network	Water contamination	Selection of improper pipe material	C	
Design	Storage	Service reservoirs	Water contamination	Inattention to proper safety measures against animal intrusion	C	Provisions for proper barricading should be considered at design phase.
Design	General	General	Interruption of water supply	Improper design due to lack and poor reliability of primary and secondary data in survey report	Water supply disruption	Survey work should be checked and validated by the designated agency as well as by the employer before work is taken up at field level.
Construction	Water source	Monitoring and surveillance	Delayed completion of the targeted project	Lack of monitoring and surveillance during construction	Water supply disruption	

Existing Control Measures (measures already in place to address hazard)	Are These Controls Effective?	Validation Notes (basis of control measure effectiveness decision)	Risk Assessment					Is Additional Control Needed?	If Yes, Proposed Controls (to be developed further in the improvement plan)
			Likelihood	Consequence	Risk Score (PxS)	Risk Level	Validation Notes		
Selection of pipe material based on prevailing Bureau of Indian Standards and Central Public Health and Environmental Engineering Organization manuals	Yes	Pipe materials selected with contamination prevention in mind	1	3	3	Low	Hazard has very low likelihood of occurrence.	No	
Property boundary wall, along with mild steel gate, considered, to prevent animal intrusion	Partly	Netting around reservoir openings should be considered to avoid animal intrusion.	2	3	6	Medium	Additional control measures should be provided to prevent animal entry through small openings.	Yes	Possible use of netting around reservoir openings to avoid animal intrusion
Third-party checking of survey report	Yes	Third-party review after survey is done by the PHED officials. But there are time and knowledge constraints.	1	1	1	Low	Minor error in secondary data	No	
Delay damage clause considered in tender document	Partly	Periodic monitoring to ensure timely completion	1	3	3	Low		Yes	Strict monitoring of implementation

continued on next page

Table A7 *continued*

Implementation Phase	Process Step	Component	Hazardous Event (X happens because of Y)		Hazard (microbial, M; physical, P; chemical, C; disruption of water supply)	Additional Information
			X = What Can Happen to the Water Supply	Y = How It Can Happen (i.e., cause)		
Construction	Treatment	Water treatment plant	Supply of untreated or contaminated water	Faulty construction of components of proposed treatment system	C	
Construction	Transport	Distribution network	Pipeline leaks and/or bursts	Technical fault due to contractor's failure to meet technical specifications for pipelines (poor joints, inadequate depth)	Water supply disruption	
Construction	Storage	Service reservoirs	Structural damage	Due to the following: - use of poor materials in construction - improper soil testing, leading to cracks	M	Previously observed in another area

Existing Control Measures (measures already in place to address hazard)	Are These Controls Effective?	Validation Notes (basis of control measure effectiveness decision)	Risk Assessment					Is Additional Control Needed?	If Yes, Proposed Controls (to be developed further in the improvement plan)
			Likelihood	Consequence	Risk Score (PxS)	Risk Level	Validation Notes		
Construction standards addressed in tender document; quality assurance and control action plans also considered	Partly	Trained personnel should be engaged to execute the work, and additional skill development program should be conducted to supplement the documentary requirement.	1	3	3	Low	Skilled personnel to be involved in implementation	Yes	Strict monitoring of implementation
Laying and jointing standards addressed in tender document; quality assurance and control action plans also considered	Partly	Trained personnel should be engaged to execute the work, and additional skill development program should be conducted to supplement the documentary requirement.	1	3	3	Low	Skilled personnel to be involved in implementation and/or maintenance	Yes	- Contractors informed by project authority about leak locations well in advance - Awareness of importance of water conservation and water waste prevention developed in the *panchayat* (village) and the community
Third-party checking of submitted soil testing report	Partly	- Standard for laying and jointing addressed in the tender document - Quality assurance and control action plans also considered	2	1	2	Low	Some cracks may be due to improper soil testing during design, causing disruption in quality and quantity of supplied water.	Yes	Selection of qualified agency with proper credentials

continued on next page

Table A7 *continued*

Implementation Phase	Process Step	Component	Hazardous Event (X happens because of Y)		Hazard (microbial, M; physical, P; chemical, C; disruption of water supply)	Additional Information
			X = What Can Happen to the Water Supply	Y = How It Can Happen (i.e., cause)		
Construction	General	General	Delay in construction	Due to the following: - lack of monitoring and surveillance during construction - bad workmanship, improper work methods, technical disputes - improper selection of agency and inadequate control over subcontracting - nonavailability of materials at construction site - local labor unrest	Water supply disruption	Labor unrest may arise from socioeconomic diversity.
Operation	Water source	Intake	Water contamination	Caused by runoff from agricultural field, wastewater discharge of towns and cities upstream of the intake location, or discharge of industrial wastes into rivers and catchment areas	C	
Operation	Water source	Intake	Water supply interruption	Improper maintenance of pumping station	Water supply disruption	Insufficient water supply to overhead reservoir

Existing Control Measures (measures already in place to address hazard)	Are These Controls Effective?	Validation Notes (basis of control measure effectiveness decision)	Risk Assessment					Is Additional Control Needed?	If Yes, Proposed Controls (to be developed further in the improvement plan)
			Likelihood	Consequence	Risk Score (PxS)	Risk Level	Validation Notes		
No control measure	No		1	2	2	Low	Additional control measures should be provided.	Yes	Work accuracy ensured by PHED and DSISC
No control measure	No		2	3	6	Medium	Presence of agricultural fields and habitations upstream of barrage and dam may lead to pesticide, microbial, etc., contamination of water.	Yes	Good agricultural practices, sensitization campaigns
Annual operation and maintenance contract	Partly		1	2	2	Low	May be due to improper maintenance of pumps	Yes	Regular joint physical verification (by contractor and PHED officials), for the smooth functioning of pumping machinery and implementation of needed repairs

continued on next page

Table A7 *continued*

Implementation Phase	Process Step	Component	Hazardous Event (X happens because of Y): X = What Can Happen to the Water Supply	Hazardous Event (X happens because of Y): Y = How It Can Happen (i.e., cause)	Hazard (microbial, M; physical, P; chemical, C; disruption of water supply)	Additional Information
Operation	Water source	Intake	Water shortage	Due to lack of rainfall in a given year	Water supply disruption	Pump operation should be regulated and low-capacity pumps should be used.
Operation	Treatment	General	Supply of untreated water	Inadequate treatment due to nonavailability, irregular supply, or low quality of chemicals	C	
Operation	Transport	Transmission main, distribution network	Water contamination	Unauthorized connections and potential backflow from illegal tapping	M	
Operation	Storage	Service reservoirs	Water contamination	Accumulation of solid particles in the reservoir floor and creation of breeding ground of algae, etc., due to lack of (or improper and sporadic) cleaning	M	

Existing Control Measures (measures already in place to address hazard)	Are These Controls Effective?	Validation Notes (basis of control measure effectiveness decision)	Risk Assessment					Is Additional Control Needed?	If Yes, Proposed Controls (to be developed further in the improvement plan)
			Likelihood	Consequence	Risk Score (PxS)	Risk Level	Validation Notes		
Supply of water should be regulated in accordance with availability of water.	Partly	Need to discuss with *gram panchayats* (village-level governance units)	2	4	8	Medium	Interaction with irrigation water department needed to ensure water conservation	Yes	- Afforestation measures - SWM-based decision support for move to another source
Steady supply of high-quality and/or branded chemicals	Partly		2	3	6	Medium	Regular supply of chemicals must be ensured, and manufacture and expiry dates confirmed.	Yes	- Chemical supplier required to follow terms and conditions of supply contract - Periodic supervision - Use of SWM to detect irregularities
Involvement of *gram panchayat;* VWSC in control of unauthorized connections.	Partly	Need to give more emphasis to the activity	4	3	12	Medium	*Gram panchayat* and/or VWSC should take necessary measures to stop unauthorized connections.	Yes	- Use of SWM to quickly detect abnormalities - Penalties for, and disconnection of, illegal connections
Regular cleaning, washing, and chlorination needed	Partly	Need to supervise maintenance	4	4	15	High		Yes	Use of surveillance camera or similar device to constantly monitor and remotely inspect such reservoirs

continued on next page

Table A7 *continued*

Implementation Phase	Process Step	Component	Hazardous Event (X happens because of Y)		Hazard (microbial, M; physical, P; chemical, C; disruption of water supply)	Additional Information
			X = What Can Happen to the Water Supply	Y = How It Can Happen (i.e., cause)		
Operation	General	General	Water contamination	Occurrence of natural disasters: flooding, earthquake, drought, etc.	M, P, C	
Operation	General	General	Water supply interruption	Labor unrest	Water supply disruption	

DSISC = design, supervision, and institutional support consultant; PHED = Public Health Engineering Department; SWM = smart water management; VWSC = village water and sanitation committee.

Source: Prepared by the consulting team, 2019.

Existing Control Measures (measures already in place to address hazard)	Are These Controls Effective?	Validation Notes (basis of control measure effectiveness decision)	Risk Assessment					Is Additional Control Needed?	If Yes, Proposed Controls (to be developed further in the improvement plan)
			Likelihood	Consequence	Risk Score (PxS)	Risk Level	Validation Notes		
- Emergency plan to be developed - Existing rules and regulations regarding disaster management in use	Partly		3	5	15	High	Natural calamities cannot be foreseen. Their occurrence could have catastrophic consequences.	Yes	- Forecasting and early warning systems - Use of mobile phone apps for quick notification of operators and consumers
Involvement of block administration and *gram panchayat* in the settlement of issues	Partly	Need to discuss with *gram panchayats*	3	3	9	Medium			- Concerned authority informed about conflict or dispute so it can be settled - Quick resolution of labor disputes - Supervision and system checking

Appendix 8: Improvement Plans

Table A8: Sample Improvement Plans

Implementation Phase	Process Step	Hazardous Event	Specific Improvement Action	Responsible Party	Estimated Budget	Funding Source	Due Date	Constraints
Design	Water source	Water supply interruption due to nonavailability of raw water caused by flaw(s) in initial bathymetric survey and unattended morphological changes like erosion or siltation, followed by inaccurate location of intake point	Bathymetric survey is SOP (specify qualified personnel requirements, equipment and calibration, secondary monitoring/supervision requirements), possibly supplemented by river sediment and morphology assessment.	Executive engineer, Planning Wing, PHED	To be developed by PHED	n.a.	As and when required	Intake point decided by irrigation department
Design	Water source	Long-term changes in water level statistics at the river intake, causing flooding, drought, or elevated salinity due to inadequate hydrological or coastal studies, and failure to consider climate change	Hydrological and salinity assessment studies, with investigation of mitigation measures	Executive engineer, Planning Wing, PHED	To be developed by PHED	n.a.	During design phase	Intake point decided by irrigation department, which must agree to any proposed changes
Design	Water treatment plant	Inadequate water treatment caused by improper design of treatment process of water due to the following: - lack of detailed assessment of raw water quality - inattention to seasonal variations in raw water quality	- Creation of PHED water quality task force to prepare guidelines for DPR water quality assessment (data compilation by water custodian authorities) and validation, water sampling, frequency and type of parameters, statistical analysis	Executive engineer, Planning Wing, PHED Executive engineer, PIU	To be developed by PHED ₹600,000	n.a. ADB	Before design phase February 2020	- DPR must be prepared within a very short period - Not enough available information about water quality - Laboratory capacity

continued on next page

Table A8 *continued*

Implementation Phase	Process Step	Hazardous Event	Specific Improvement Action	Responsible Party	Estimated Budget	Funding Source	Due Date	Constraints
	Water treatment plant		- In year 1, collection of six water samples (according to season: summer, rainy season/ monsoon, autumn, pre-winter, winter, spring) - Parameters: all IS 10500 (2012) parameters prescribed for inland surface water quality by the Ministry of Environment, Forests and Climate Change					
Construction	Service reservoirs	Structural damage due to the following: - use of poor materials in construction - Improper soil testing may lead to cracks	- Development and implementation of material and testing verification procedures to be carried out at field level by PIU - Formulation of proper guidelines for unbiased selection of vendor with sufficient production capacity and proper QAP	PMU, PIU, and PMC	To be developed by PHED	n.a.	Before construction	
Construction	General	Delay in construction phase due to - lack of monitoring and surveillance during construction - poor workmanship, improper work methods, technical disputes	- PIU development and implementation of random checking procedure at field level - Revision of tendering process or strict supervision and imposition of relevant penalty clause	DSISC and PIU	To be developed by PHED	n.a.	Before construction	

continued on next page

Table A8 *continued*

Implementation Phase	Process Step	Hazardous Event	Specific Improvement Action	Responsible Party	Estimated Budget	Funding Source	Due Date	Constraints
	General	- improper selection of agency and inadequate control over subcontracting - nonavailability of materials at construction site - local labor unrest						
Operation	General	Supply of untreated and/or contaminated water caused by inadequate water treatment due to lack of proper technical knowledge of operators	- Training program for electromechanical personnel - Development of IEC plan and training materials for awareness generation and skills development of operators	PHED				
Operation	General	Water supply interruption due to power failure	- Development of SOPs for prompt response during power cuts and similar situations - Training for technical personnel	DSISC and PIU, PHED	n.a.; to be developed by PHED	n.a.	March 2024	
Operation	Water intake	Water supply interruption due to elevated salinity levels	Development of SOPs for scheduling of pumping and use of buffer storage reservoirs	Executing agency, DSISC, and PIU, PHED	n.a.	n.a.	During design phase	

continued on next page

Table A8 *continued*

Implementation Phase	Process Step	Hazardous Event	Specific Improvement Action	Responsible Party	Estimated Budget	Funding Source	Due Date	Constraints
Operation	General	Supply of untreated water caused by inadequate treatment due to nonavailability or low quality and irregular supply of chemicals	- Measures to ensure regular supply of chemicals, confirming dates of manufacturing and expiry - Supplier should follow the terms and conditions of chemical supply contract	DSISC and PIU, PHED	n.a.; to be developed by PHED	n.a.	March 2024	
Operation	Disinfection	Supply of untreated and/or inadequately treated water due to unmonitored presence of residual chlorine in the water	- Activation of existing VWSC/VHSNC for spot testing for residual chlorine, with adequate training support and report submission requirements - Over time, online real-time residual chlorine monitoring through SWM, to supplement spot testing	PIU/PHED and P&RDD	Around ₹500,000/ per year	NRDWP support fund	March 2024	
Operation	Transmission main, distribution network	Water supply interruption caused by lack of coordination between contractors and operators	- Regular communication with district and/or block administrators and *gram panchayat* (village-level governance unit) or VWSC to keep them informed of disputes - Regular communication with agencies and/or contractors engaged for O&M	Executing agency (contractor), PIU/PHED, P&RDD, and district administration	n.a.	n.a.	When necessary	

continued on next page

Table A8 *continued*

Implementation Phase	Process Step	Hazardous Event	Specific Improvement Action	Responsible Party	Estimated Budget	Funding Source	Due Date	Constraints
Operation	Transmission main, distribution network	Water supply interruption caused by pipe leaks and/or bursts	- Skill development training to facilitate prompt and sanitary repair work - Over time, online real-time pressure monitoring and flow monitoring through SWM, to provide early warning of pipe leaks and/or bursts	Executing agency (contractor) and PIU, PHED	n.a.; to be developed by PHED	n.a.	March 2024	
Operation	Transmission main, distribution network	Water contamination caused by unauthorized connections and potential backflow from illegal tapping	- Firm handling of this issue by the *panchayat* (village) or local administration - Necessary action by *gram panchayat*/VWSC to stop unauthorized connections - Imposition of penalty	NGO; DSISC; PIU, PHED; and P&RDD	n.a.	n.a.	When necessary	
Operation	Transmission main, distribution network	Water contamination following biofilm formation due to lack of chlorine	Development of online system for monitoring chemical stock	Project director, PMU, PHED	n.a.; to be developed by PHED	n.a.	March 2024	
Operation	Service reservoirs	Water contamination caused by accumulation of solid particles in the reservoir floor and creation of breeding ground for algae, etc., due to lack of, or improper and periodic, cleaning	Proper supervision and monitoring of activities related to cleaning of reservoir, etc.	PHED, contractor	n.a.	n.a.	March 2024	

continued on next page

Table A8 *continued*

Implementation Phase	Process Step	Hazardous Event	Specific Improvement Action	Responsible Party	Estimated Budget	Funding Source	Due Date	Constraints
Operation	General	Water supply interruption caused by nonavailability of spare parts in time of unpredictable crisis	- Management of a spare parts inventory to enable prompt response to crisis - Over time, establishment of a computerized maintenance management system and GIS-based asset management in SWM, to provide full transparency regarding availability of spare parts		To be determined	n.a.	March 2024	

₹ = Indian rupee, ADB = Asian Development Bank; DPR = detailed project report; DSISC = design, supervision, and institutional support consultant; GIS = geographic information system; IEC = information, education, and communication; n.a. = not applicable; NGO = nongovernment organization; NRDWP = National Rural Drinking Water Program; O&M = operation and maintenance; P&RDD = Panchayat & Rural Development Department; PHED = Public Health Engineering Department; PIU = project implementation unit; PMC = project management consultant; PMU = project management unit; QAP = quality assurance plan; SOP = standard operating procedure; SWM = smart water management; VHSNC = village health, sanitation, and nutrition committee; VWSC = village water and sanitation committee.

Source: Prepared by the consulting team, 2019.

Appendix 9: Operational Monitoring of Control Measures

Table A9: Sample Operational Monitoring of Control Measures

Implementation Phase	Process Step	Hazardous Event	Control Measure(s)	Monitoring			
				Operational Limit (or target condition)	What Should Be Monitored?	Where?	When?
Design	Water source	Long-term changes in water-level statistics at the river intake, causing elevated salinity, with the need for intelligent pump management or provision of alternative supplementary water sources	Water intake design, real-time monitoring, and increase in buffer storage capacity, or use of alternative water source	Optimum design, considering frequency of elevated salinity levels, technical robustness, and economics	Project design	Planning Division	During design period
Design	Water treatment plant	Inadequate water treatment caused by improper design of the water treatment process due to the following: - lack of detailed assessment of raw water quality - inattention to seasonal variations in raw water quality	Treatment plant design to be based on available raw water quality data and survey data to effectively remove all contaminants; conventional type of water treatment plant proposed	Optimum design in relation to standards of the Central Public Health & Environmental Engineering Organisation (CPHEEO) and the Ministry of Urban Development	Project design	Planning Division	Before inviting bids
Design	Pumping station	Water supply interruption caused by malfunctioning of pumping station due to wrong selection of capacity and number of pumps	Proof checking to be vetted by third party (DSISC) and approved by PHED	Optimum design in relation to standards of the CPHEEO and the Ministry of Urban Development	Project design	Planning Division	Before inviting bids
Design	Transmission main, distribution network	Water contamination caused by the selection of improper pipe material	Conformity of pipe material thoroughly checked during procurement and pipe material selected on the	Pipe material selected considering all prevailing manuals and guidelines	Project design	Office	Before finalizing bid document

Monitoring *(continued)*				Corrective Action(s)					
	How?	Who Should Monitor?	What Records Should Be Kept?	What Action(s) Should Be Taken?	When Should It/They Be Implemented?	Who Should Implement the Action(s)?	Who Should Be Informed?	What Procedure Should Be Followed?	What Records Should Be Kept?
	Record checking	Consultants (third party)	Design records	Modification if necessary, according to findings	During design, by contractor	Contractor	PIU and PMU	Approval of design	Submitted design
	Record checking	Consultants (third party)	Design records	Modification if necessary, according to findings	During DPR preparation and before inviting bids	Planning Division	Responsible officials of the Planning Division		
	Record checking	Consultants (third party)	Design records	Modification if necessary, according to findings	During DPR preparation and before inviting bids	Planning Division	Responsible officials of the Planning Division		
	According to quality control and quality assurance system guidelines	PHED	Official documen-tation	Modification of documents to incorporate PHED comments and/or observations	Immediately	Consultant	PIU	Rechecking of compliance with prescribed guidelines	Official documents

continued on next page

Table A9 *continued*

Implementation Phase	Process Step	Hazardous Event	Control Measure(s)	Monitoring			
				Operational Limit (or target condition)	What Should Be Monitored?	Where?	When?
	Transmission main, distribution network		basis of prevailing Bureau of Indian Standards (BIS) and CPHEEO guidelines				
Design	General	Water supply interruption caused by improper design due to lack or poor reliability of primary and secondary data in survey report	Implementation of design and/or project technical audit (proof checking)	Conformity with the CPHEEO and the BIS Code	Design criteria	Design documents	Before inviting bids
Construction	Water treatment plant	Supply of untreated or contaminated water caused by faulty construction of components of proposed treatment system	- Standard requirements of the project specified in bid document - Skilled personnel to be involved in execution - Strict monitoring should be done during the execution.	Compliance with guidelines	Quality control during execution	Site	During construction period
Construction	Distribution network	Pipe leaks and/or bursts caused by technical fault due to failure of contractor to meet technical specifications for pipelines (poor joints, inadequate depth)	- Pipe laying and jointing standards addressed in tender document - Quality assurance and control action plans also considered	Compliance with guidelines	Pipe laying and jointing procedure	Site	During construction

Monitoring *(continued)*				Corrective Action(s)					
	How?	Who Should Monitor?	What Records Should Be Kept?	What Action(s) Should Be Taken?	When Should It/They Be Implemented?	Who Should Implement the Action(s)?	Who Should Be Informed?	What Procedure Should Be Followed?	What Records Should Be Kept?
	Through scrutiny	PIU and DSISC	Approved by the DPR	Review of design	Before inviting bids	PIU and DSISC	PMU		Submitted revised DPR
	Through strict monitoring of execution	PHED and DSISC	Official documents	Rejection of construction and submission of request for new construction	Immediately	Contractor	PIU and DSISC	Rechecking of compliance with prescribed guidelines	Site order book
	Through strict monitoring	PIU and DSISC	Site order book	Rejection of faulty pipeline and request addressed to contractor to relay the pipeline	Immediately	Contractor	PIU and DSISC	Compliance with specifications	Site order book

continued on next page

Table A9 *continued*

Implementation Phase	Process Step	Hazardous Event	Control Measure(s)	Monitoring			
				Operational Limit (or target condition)	What Should Be Monitored?	Where?	When?
Construction	Service reservoirs	Structural damage due to the following: - use of poor materials in construction - improper soil testing, leading to cracks	- Third-party checking of submitted soil-testing report - Selection of proper source of different materials according to IS 383 and other codal provisions and high-quality monitoring of site-level testing - Leakage testing in the field after completion of entire work according to IS 3370-4 (1967) and other relevant standards	Sufficient contractor credentials according to contract terms and conditions	Technical papers of bidder and soil test report according to IS code		Before inviting bids
Construction	General	Delay in construction phase due to the following: - lack of monitoring and surveillance during construction - poor workmanship, improper work methods, technical disputes - improper selection of agency and inadequate control over subcontracting - Nonavailability of material at construction site - Local labor unrest	- Periodic monitoring to ensure progress according to time schedule - Possible inclusion of delay damage clause in tender document	Conformity with contract documents	Documents submitted by contractor in relation to sub-contracting	Office	Before taking up the construction work

Monitoring *(continued)*				Corrective Action(s)					
	How?	Who Should Monitor?	What Records Should Be Kept?	What Action(s) Should Be Taken?	When Should It/They Be Implemented?	Who Should Implement the Action(s)?	Who Should Be Informed?	What Procedure Should Be Followed?	What Records Should Be Kept?
	- Thorough scrutiny - Review of credentials of agency according to bid stipulation during technical evaluation - Checking should be done in relation to IS code.	PIU and DSISC	Proof checking	Modification of guidelines from time to time to conform to latest version of IS codes and contract clause	Before approval	PIU and DSISC	PMU	Proper performance evaluation	Procedural documents
	Checking of criteria	PIU and DSISC	Issuance of letter for approval of sub-contractor						

continued on next page

Table A9 *continued*

Implementation Phase	Process Step	Hazardous Event	Control Measure(s)	Monitoring			
				Operational Limit (or target condition)	What Should Be Monitored?	Where?	When?
Construction	General	Water contamination caused by improper cleaning of reservoirs and transmission mains before commissioning	Provision kept in the tender document and works to be properly supervised	Compliance with guidelines	Strict monitoring should be done by DSISC and PHED so that safe water is supplied regularly	Pipeline	Before commissioning
Operation	Intake	Water supply interruption caused by irregular pumping and/or supply of raw water due to physical damage in intake well	- Regular cleaning of suction pipe screen at intake and liaison with river authority - Ganga Action Plan (Namami Gange Project)	No blockage by aquatic weeds	Weed and/or algae density	Intake, WTP	Periodic monitoring according to O&M manual
Operation	Intake	Long-term changes in water-level statistics at the river intake causing elevated salinity, with the need for intelligent SWM of pumping	Online measurement of salinity at the intake; documentation of compliance with drinking water standards	<500 mg/l	TDS as a measure of salinity	Intake site	- Continously and online - Pointwise control measurement by PHED
Operation	General	Interruption of water supply due to power failure	Alternative back-up power facility is available	Generator is working	Generator starts working when power supply fails	Generator	Every 3 months
Operation	General	Water contamination and water supply interruption caused by the occurrence of natural disasters: flooding, earthquake, drought, etc.	- Development of emergency plan - Existing rules and regulations regarding disaster management are in use	Compliance with emergency procedures	Staff compliance with emergency procedures	To be specified	Once a year

BIS = Bureau of Indian Standards; CPHEEO = Central Public Health & Environmental Engineering Organisation; DPR = detailed project report; DSISC = design, supervision, and institutional support consultant; mg/l = milligram per liter; O&M = operation and maintenance; PHED = Public Health Engineering Department; PIU = project implementation unit; PMU = project management unit; SCADA = supervisory control and data acquisition; SOP = standard operating procedure; SWM = smart water management; TDS = total dissolved solid; WTP = water treatment plant.

Source: Prepared by the consulting team, 2019.

Monitoring *(continued)*				Corrective Action(s)					
	How?	Who Should Monitor?	What Records Should Be Kept?	What Action(s) Should Be Taken?	When Should It/They Be Implemented?	Who Should Implement the Action(s)?	Who Should Be Informed?	What Procedure Should Be Followed?	What Records Should Be Kept?
	By testing end-point water quality	PHED and DSISC	Lab Test Registrar	Issuance of letter to executing agency (contractor) requesting execution of work	Immediately	Contractor	PIU and DSISC	Official procedures	Official documents
	Visual	Ground staff members of authorized agency and/or agencies	Logbook and/or O&M app	Mechanical weed removal process in large scale	Whenever clogging is noticed	Ground staff members of implementing agency	Operation manager and authorized agency and/or agencies	Mechanical weed removal process in large scale	Logbooks to be monitored for ground staff
	Sensor, linked to SCADA system	Contractor	Data server	SOP updating if not in compliance	Commissioning of system	Contractor	Executive engineer (PHED)	Change in management of monitoring system	Data server
	Visual checking	PHED (division to be defined) Executive engineer and/or representative	Logbook and/or designated online app	Repair and checking to make sure that generator is working, and that supply of petrol and diesel is adequate	Immediately	Contractor	Executive engineer (PHED)		Logbooks
	Observation	Representative of Executive engineer of PHED and/or disaster management representative	Logbook and/or designated online app	Training in emergency procedures	Agreement between parties (PHED and contractors)	Contractor	Executive engineer (PHED)	Emergency plan	Training record

Students of Safique Ahmed Girls High School in their classroom. Haroa, 24 North Pargana, West Bengal, India.

REFERENCES

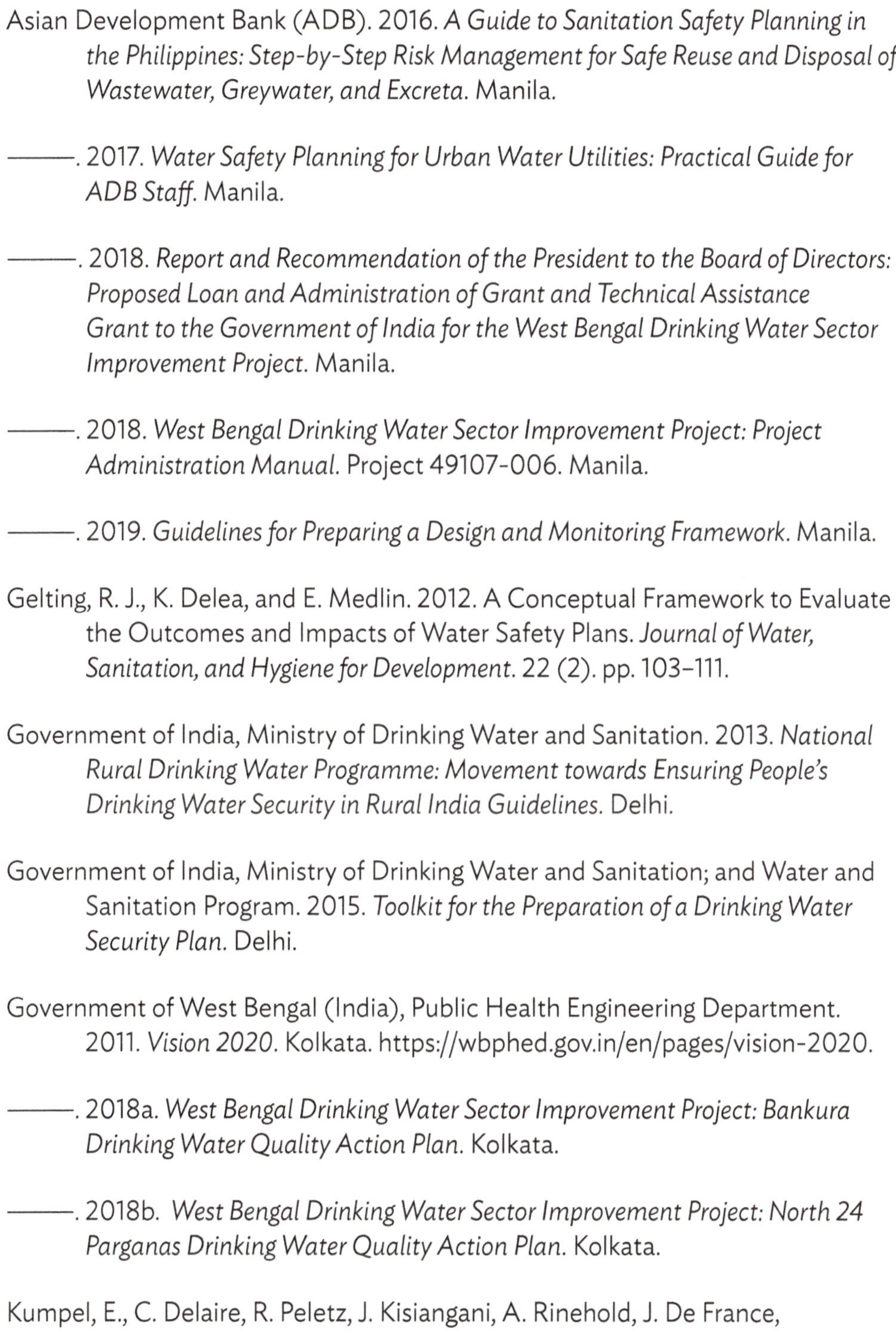

Asian Development Bank (ADB). 2016. *A Guide to Sanitation Safety Planning in the Philippines: Step-by-Step Risk Management for Safe Reuse and Disposal of Wastewater, Greywater, and Excreta.* Manila.

———. 2017. *Water Safety Planning for Urban Water Utilities: Practical Guide for ADB Staff.* Manila.

———. 2018. *Report and Recommendation of the President to the Board of Directors: Proposed Loan and Administration of Grant and Technical Assistance Grant to the Government of India for the West Bengal Drinking Water Sector Improvement Project.* Manila.

———. 2018. *West Bengal Drinking Water Sector Improvement Project: Project Administration Manual.* Project 49107-006. Manila.

———. 2019. *Guidelines for Preparing a Design and Monitoring Framework.* Manila.

Gelting, R. J., K. Delea, and E. Medlin. 2012. A Conceptual Framework to Evaluate the Outcomes and Impacts of Water Safety Plans. *Journal of Water, Sanitation, and Hygiene for Development.* 22 (2). pp. 103–111.

Government of India, Ministry of Drinking Water and Sanitation. 2013. *National Rural Drinking Water Programme: Movement towards Ensuring People's Drinking Water Security in Rural India Guidelines.* Delhi.

Government of India, Ministry of Drinking Water and Sanitation; and Water and Sanitation Program. 2015. *Toolkit for the Preparation of a Drinking Water Security Plan.* Delhi.

Government of West Bengal (India), Public Health Engineering Department. 2011. *Vision 2020.* Kolkata. https://wbphed.gov.in/en/pages/vision-2020.

———. 2018a. *West Bengal Drinking Water Sector Improvement Project: Bankura Drinking Water Quality Action Plan.* Kolkata.

———. 2018b. *West Bengal Drinking Water Sector Improvement Project: North 24 Parganas Drinking Water Quality Action Plan.* Kolkata.

Kumpel, E., C. Delaire, R. Peletz, J. Kisiangani, A. Rinehold, J. De France, D. Sutherland, and R. Khush. 2018. Measuring the Impacts of Water Safety Plans in the Asia-Pacific Region. *International Journal of Environmental Research and Public Health.* 15 (6). pp. 12–23. doi: 10.3390/ijerph15061223.

Lockhart, G., W. E. Oswald, B. Hubbard, E. Medlin, and R. J. Gelting. 2014. Development of Indicators for Measuring Outcomes of Water Safety Plans. *Journal of Water, Sanitation, and Hygiene for Development*. 4 (1). pp. 171–181. doi: 10.2166/washdev.2013.159.

Samwel, M. and C. Wendland, eds. 2016. *Developing a Water & Sanitation Safety Plan in a Rural Community: How to Accomplish a Water and Sanitation Safety Plan? Compendium – Part A*. 2nd revised ed. Utrecht, The Netherlands: Women in Europe for a Common Future (WECF International). http://www.wecf.org/wp-content/uploads/2018/11/WSSPPublicationENPartA.pdf.

Sulabh International Academy of Environmental Sanitation. 2009. *Guidelines for Water Safety Plans for Rural Water Supply Systems*. With support from the World Health Organization India Country Office. Delhi.

Water and Sanitation Program. 2010. *Water Safety Plans for Rural Water Supply in India: Policy Issues and Institutional Arrangements*. Delhi.

World Health Organization (WHO). 2012. *Water Safety Planning for Small Community Water Supplies: Step-by-Step Risk Management Guidance for Drinking-Water Supplies in Small Communities*. Geneva. https://www.who.int/water_sanitation_health/publications/small-comm-water_supplies/en/.

———. 2015. *Sanitation Safety Planning: Manual for Safe Use and Disposal of Wastewater, Greywater and Excreta*. Geneva. https://www.who.int/water_sanitation_health/publications/ssp-manual/en/.

———. 2017. *Guidelines for Drinking-Water Quality*. 4th ed. Geneva. https://www.who.int/water_sanitation_health/publications/drinking-water-quality-guidelines-4-including-1st-addendum/en/.

WHO South-East Asian Regional Office (SEARO). 2016. *Capacity Training on Urban Water Safety Planning: Training Modules*. Delhi.

WHO Western Pacific Regional Office (WPRO). 2008. *Training Workbook on Water Safety Plans for Urban Systems*. Manila.

WHO and International Water Association (IWA). 2009. *Water Safety Plan Manual: Step-by-Step Risk Management for Drinking-Water Suppliers*. Geneva.

———. 2015. *A Practical Guide to Auditing Water Safety Plans*. Geneva.

www.ingramcontent.com/pod-product-compliance
Lightning Source LLC
Chambersburg PA
CBHW050044220326
41599CB00045B/7272

* 9 7 8 9 2 9 2 6 2 5 2 7 6 *